Geospatial Technology

Geospatial Technology
Fundamentals and Applications

By

V. Emayavaramban
K. Kannadasan
S. Vinothkanna

NEW INDIA PUBLISHING AGENCY
New Delhi – 110 034

NEW INDIA PUBLISHING AGENCY
101, Vikas Surya Plaza, CU Block, LSC Market
Pitam Pura, New Delhi 110 034, India
Phone: + 91 (11)27 34 17 17 Fax: + 91(11) 27 34 16 16
Email: info@nipabooks.com
Web: www.nipabooks.com

Feedback at feedbacks@nipabooks.com

ISBN 978-93-85516-80-1

Composed, Designed by NIPA

Preface

The Geospatial technology is a term used to describe the range of modern tools contributing to the geographic mapping and analysis of the Earth environment and human activities. These technologies have been evolving in some form since the first maps were drawn in prehistoric times. In days all kinds of activities and analysis may included with showing maps.

"Technology relating to the collection or processing of data that is associated with location"

Satellites allowed images of the Earth's surface and human activities therein with certain limitations. Computers allowed storage and transfer of imagery together with the development of associated digital software, maps, and data sets on socioeconomic and environmental phenomena, collectively called geographic information systems (GIS). An important aspect of a GIS is its ability to assemble the range of geospatial data into a layered set of maps which allow complex themes to be analyzed and then communicated to wider audiences. This 'layering' is enabled by the fact that all such data includes information on its precise location on the surface of the Earth, hence the term 'geospatial'.

Especially in the last decade, these technologies have evolved into a network of national security, scientific, and commercially operated satellites complemented by powerful desktop GIS.

High quality hardware and data is now available to new audiences such as universities, corporations, and non-governmental organizations. The fields and sectors deploying these technologies are currently growing at a rapid pace, informing decision makers on topics such as industrial engineering, biodiversity conservation, forest fire suppression, agricultural monitoring, humanitarian relief, and much more.

A *Geospatial network* is a network of collaborating resources for sharing and coordinating geographical data and data tied to geographical references.

Geomatics is a field that incorporates several others such as the older field of land surveying engineering along with many other aspects of spatial data management ranging from data science and cartography to geography. Following the advanced developments in digital data processing, the nature of the tasks required of the professional land surveyor has evolved and the term "surveying" no longer accurately covers the whole range of tasks that the profession deals with. Like how the profession of mechanics is a part of Mechanical engineering, surveying is a part of Geomatics engineering.

Geospatial technology refers to equipment used in visualization, measurement, and analysis of earth's features, typically involving such systems as GPS (global positioning systems), GIS (Geographical Information Systems), and RS (Remote Sensing). Its use is well-known and widespread in the military and in homeland security, but its influence is pervasive everywhere, even in areas with a lower public profile, such as land use, flood plain mapping and environmental protection.

The book has been divided into six chapters. The first chapter "Introduction to remote sensing" in this chapter we have given basics knowledge of remote sensing, it include the definition, history of the satellite systems, electromagnetic spectrum and resolutions etc. The second chapter "Aerial Remote Sensing" deals with elements, interpretation, types of photo and measurements. The third chapter represents the "Satellite Remote Sensing". The fourth chapter analysed "Global Positioning System", it include the GPS procedure, types and developments. The fifth chapter is "Introduction to Geographical Information System" and covered basics information about GIS. The sixth chapter include applications of remote sensing, geographical information system and Global positioning systems.

This book is very useful to bachelor, post graduates' degree of geography and research scholars can get the some ideas and knowledge about the Geospatial Technologies. This book is prepared with very basic and common information.

Geographic knowledge of these technologies will provide students with the skills needed to meet the demands of their diverse careers in the global workplace.

V. Emayavaramban
K. Kannadasan
S. Vinothkanna

Contents

1

Introduction to Remote Sensing

Introduction

Remote sensing is the art and science of recording, measuring, and analyzing information about a phenomenon from a distance. Humans with the aid of their eyes, noses, and ears are constantly seeing, smelling, and hearing things from a distance as they move through an environment. Thus, humans are naturally designed to be remote sensors. In order to study large areas of the Earth's surface geographers use devices known as remote sensors. These sensors are mounted on platforms such as helicopters, planes, and satellites that make it possible for the sensors to observe the Earth from above.

Definition

Remote sensing can be defined as any process whereby information is gathered about an object, area or phenomenon without being in contact with it. Our eyes are an excellent example of a remote sensing device. We are able to gather information about our surroundings by gauging the amount and nature of the reflectance of visible light energy from some external source (such as the sun or a light bulb) as it reflects off objects in our field of view. Contrast this with a thermometer, which must be in contact with the phenomenon it measures, and thus is not a remote sensing device.

Given this rather general definition, the term *remote sensing* has come to be associated more specifically with the gauging of interactions between earth surface materials and electromagnetic energy. However, any such attempt at a more specific definition becomes difficult, since it is not always the natural environment that is sensed (e.g., art conservation applications), the energy type is not always electromagnetic (e.g., sonar) and some procedures gauge natural energy emissions (e.g., thermal infrared) rather than interactions with energy from an independent source.

History of Remote Sensing

The technology of modern remote sensing began with the invention of the camera more than 150 years ago. Although the first, rather primitive photographs were taken as "stills" on the ground, the idea and practice of looking down at the Earth's surface emerged in the 1840s when pictures were taken from cameras secured to tethered balloons for purposes of topographic mapping. Perhaps the most novel platform at the end of the last century is the famed pigeon fleet that operated as a novelty in Europe. By the first World War, cameras mounted on airplanes provided aerial views of fairly large surface areas that proved invaluable in military reconnaissance. From then until the early 1960s, the aerial photograph remained the single standard tool for depicting the surface from a vertical or oblique perspective.

Satellite remote sensing can be traced to the early days of the space age (both Russian and American programs) and actually began as a dual approach to imaging surfaces using several types of sensors from spacecraft. In 1946, V-2 rockets acquired from Germany after World War II were launched to high altitudes from White Sands, New Mexico. These rockets, while never attaining orbit, contained automated still or movie cameras that took pictures as the vehicle ascended. Then, with the emergence of the space program in the 1960s, Earth-orbiting cosmonauts and astronauts acted much like tourists by taking photos out the window of their spacecraft.

The term "remote sensing," first used in the United States in the 1950s by Ms. Evelyn Pruitt of the U.S. Office of Naval Research, is now commonly used to describe the science—and art—of identifying, observing, and measuring an object without coming into direct contact with it. This process involves the detection and measurement of radiation of different wavelengths reflected or emitted from distant objects or materials, by which they may be identified and categorized by class/type, substance, and spatial distribution.

The period from 1960 to 2010 has experienced some major changes in the field of remote sensing. The background for many of these changes occurred in the 1960s and 1970s. Some of these changes are outlined below.

- First, the term "remote sensing" was initially introduced in 1960. Before 1960 the term used was generally aerial photography. However, new methods and technologies for sensing of the Earth's surface were moving beyond the traditional black and white aerial photograph, requiring a new, more comprehensive term be established.

- Second, the 1960s and 1970s saw the primary platform used to carry remotely sensed instruments shift from air planes to satellites. Satellites can cover much more land space than planes and can monitor areas on a regular basis.
- Third, imagery became digital in format rather than analog. The digital format made it possible to display and analyze imagery using computers, a technology that was also undergoing rapid change during this period. Computer technology was moving from large mainframe machines to small microcomputers and providing information more in graphic form rather than numerical output.
- Fourth, sensors were becoming available that recorded the Earth's surface simultaneously in several different portions of the electro-magnetic spectrum. One could now view an area by looking at several different images, some in portions of the spectrum beyond what the human eye could view. This technology made it possible to see things occurring on the Earth's surface that looking at a normal aerial photograph one could not detect.
- Finally, the turbulent social movements of the 1960s and 1970s awakened a new and continuing concern about the changes in the Earth's physical environment. Remotely sensed imagery from satellites - analyzed and enhanced with computers - made it possible to detect and monitor these changes. Thus, societal support was and continues to remain strong for this technology, even though very few people are familiar with the term, remote sensing.

Today, many satellites, with various remote sensing instruments, monitor the Earth's surface. These satellites and their respective remote sensing programs can trace their origins back to the CORONA and Landsat programs. CORONA was a secretive military reconnaissance program that continues to the present time through the advanced Keyhole satellites and Landsat was an open Earth resources program that also continues through more advanced Landsats and other satellite resource monitoring programs. This unit centers on the development and growth of these two programs.

CORONA : America's First Satellite Program

The CORONA Program occurred between 1959 and 1972 but the American public did not know about the project until 1995 when President Bill Clinton ordered the declassification of the imagery. Vice President Al Gore urged the declassification of the imagery in order to assist scientists in conducting various environmental studies.

U-2 Aircraft Reconnaissance Program

To appreciate the importance of the CORONA Program some historical context is required. After World War II and through the 1950s, the Cold War with the Soviet Union created a need for the United States to know about the military strength of the Soviet Union, especially with respect to the number of long range bombers and intercontinental ballistic missiles (ICBMs) that it possessed. Reconnaissance planes were being sent across the borders into the Soviet Union and its allied nations. However, these planes could only penetrate the edges of the country and the information needed was in the middle of the Soviet Union's large land mass. Many of these planes and their crews never returned. A new plane was required that could fly high enough that Soviet planes and surface-to-air missiles could not shoot it down. More importantly this plane needed to be able to take detailed aerial photographs over the middle of the Soviet Union.

In 1954, President Dwight Eisenhower had the CIA, working in conjunction with the U.S. Air Force, negotiate an arrangement with Lockheed to create such an airplane. This arrangement led to the development of the secret Lockheed Skunk Works where Kelly Johnson and his engineers created within a few months the U-2. Eisenhower realized that the use of such a plane over the territory of another nation was in violation of the other nation's sovereignty and could result in military reprisal. However, he was under pressure from the military-industrial complex in the United States to address the supposed large bomber and ICBM advantage that the Soviet Union had over the U.S. and the verbal threats being made by the Soviet leadership. Before using such a plane he proposed to the Soviet Union that both countries be allowed to send reconnaissance planes across each other's territory in order to permit each country to have accurate information about the other country's military strength. This proposal was rejected by the Soviet Union.

The first overflight of the Soviet Union by a U-2 occurred on July 4, 1956. It took off from West Germany and flew over Poland, Belorussia, Moscow, Leningrad, and the Soviet Baltic states. It flew at 70,000 feet, about 25,000 feet higher than any other plane could fly at the time, and exposed 3,600 feet of film per mission using two high-resolution cameras. One camera had a long focal length allowing it to spot items that were two to three feet across. These flights occurred several times over the next few years followed each time by formal protests by the Soviet Union. U-2 flights were also made over other nations. Finally, on May 1, 1960, using a cluster of 14 SA-2 missiles, the Soviet Union was able to shoot down a U-2 over its territory. The plane's pilot, Francis Gary Powers, survived, and in August 1960 the Soviets staged a highly publicized trial to embarrass the United States. The United States agreed never to fly a reconnaissance plane across the Soviet Union again. This agreement included Communist China and Eastern Europe. These countries became known as the "denied areas." The U-2 flights over the Soviet Union showed that no bomber or missile gap existed. Based on this information Eisenhower, an ex-military man, warned the nation immediately before leaving the presidency about the influence of the military-industrial complex and its role in shaping U.S. policy. It was the pressure from this complex that lead to the development of the U-2 and the possible conflict with the Soviet Union. The U-2 was later used in other sections of the world, most noticeably Cuba and Vietnam.

The number of bombers can be counted as they are arranged along the length of the field. Several open spaces exist along the alignment suggesting parking areas for planes that might be in flight. Two bombers are located in the upper left side of the photo and can easily taxi to the runway for a quick takeoff. A bomber also appears to be located in the midfield. At both ends of the airfield are smaller planes, probably fighters to protect the airfield from attack. The bombers are most likely the Tupolev Tu-95, a large, four-engine turboprop powered plane that was both a long range bomber and missile platform. Its distinctively swept back wings are at 35 degrees, a very sharp angle by the standards of propeller-driven aircraft.

Passive and Active Sensor Systems

Two types of sensors exist, namely passive and active. A passive sensor system needs an external energy source (Figure1.1). In most cases this source is the sun. These sensors generally detect reflected and emitted energy wave lengths from a phenomenon. An active sensor system provides its own energy source. As an example, a radar sensor sends out sound waves and records the reflection waves coming back from the surface. Passive systems are much more common than active systems.

Most sensors record information about the Earth's surface by measuring the transmission of energy from the surface in different portions of the electromagnetic (EM) spectrum (Figure 1.2). Because the Earth's surface varies in nature, the transmitted energy also varies. This variation in energy allows images of the surface to be created. Human eyes see this variation in energy in the visible portion of the EM spectrum. Sensors detect variations in energy in both the visible and non-visible areas of the spectrum.

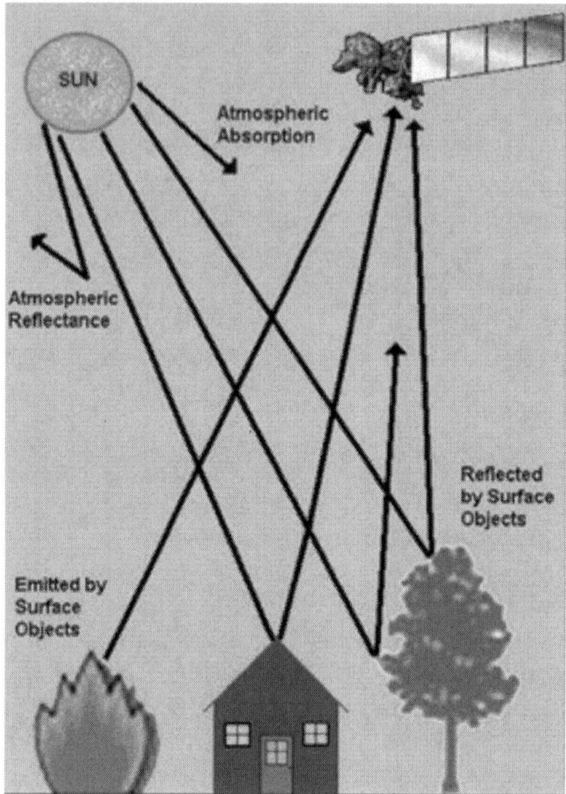

Figure 1.1. Passive System
(See colour version on page 169)

Energy waves in certain sections of the EM spectrum easily pass through the atmosphere, while other types do not. The ability of the atmosphere to allow energy to pass through it is referred to as its transmissivity, and varies with the wavelength/type of the radiation. The gases that comprise our atmosphere absorb energy in certain wavelengths while allowing energy with differing wavelengths to pass through.

Electro Magnetic Spectrum

The areas of the EM spectrum that are absorbed by atmospheric gases such as water vapor, carbon dioxide, and ozone are known as absorption bands. In Figure 1.2, absorption bands (shown in brown) are represented by a low transmission value that is associated with a specific range of wavelengths. Trying to obtain remotely sensed imagery in the absorption bands is nearly impossible; thus, sensors are generally designed not to record information in these portions of the spectrum.

In contrast to the absorption bands, there are areas of the EM spectrum, where the atmosphere is transparent (little or no absorption of energy) to specific wavelengths. These wavelength bands are known as atmospheric "windows" since they allow the energy to easily pass through the atmosphere to Earth's surface. It is in these windows that sensors are used to gather information about Earth phenomena.

Most remote sensing instruments on aircraft or space-based platforms operate in one or more of these windows by making their measurements with detectors tuned to specific frequencies (wavelengths) that pass through the atmosphere. When a remote sensing instrument has a line-of-sight with an object that is reflecting sunlight or emitting heat, the instrument collects and records the radiant energy. While most remote sensing systems are designed to collect reflected energy, some sensors, especially those on meteorological satellites, directly measure absorption phenomena, such as those associated with carbon dioxide (CO_2) and other gases. The atmosphere is nearly opaque to EM energy in part of the mid-IR and all of the far-IR regions. In the microwave region, by contrast, most of this radiation moves through unimpeded, so radar waves reach the surface (although weather radars are able to detect clouds and precipitation because they are tuned to observe backscattered radiation from liquid and ice particles).

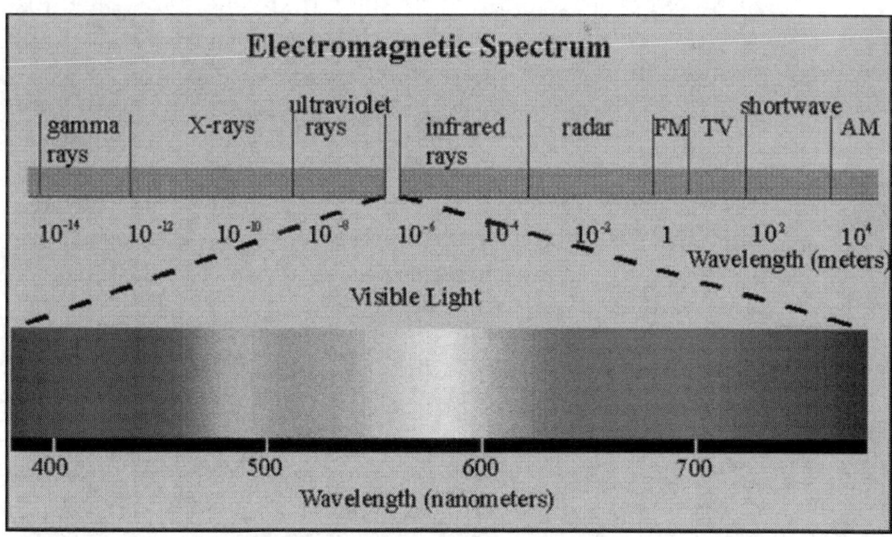

Figure 1.2. Electromagnetic (EM) Spectrum.
(See colour version on page 169)

Atmospheric Windows

The areas of the EM spectrum that are absorbed by atmospheric gases such as water vapor, carbon dioxide, and ozone are known as absorption bands. The absorption bands (shown in brown) are represented by a low transmission value that is associated with a specific range of wavelengths. Trying to obtain remotely sensed imagery in the absorption bands is nearly impossible; thus, sensors are generally designed not to record information in these portions of the spectrum.

In contrast to the absorption bands, there are areas of the EM spectrum (shown in green in Figure 1.2 and Table 1.1) where the atmosphere is transparent (little or no absorption of energy) to specific wavelengths. These wavelength bands are known as atmospheric "windows" since they allow the energy to easily pass through the atmosphere to Earth's surface. It is in these windows that sensors are used to gather information about Earth phenomena.

Most remote sensing instruments on aircraft or space-based platforms operate in one or more of these windows by making their measurements with detectors tuned to specific frequencies (wavelengths) that pass through the atmosphere. When a remote sensing instrument has a line-of-sight with an object that is reflecting sunlight or emitting heat, the instrument collects and records the radiant energy. While most remote sensing systems are designed to collect reflected energy, some sensors, especially those on meteorological satellites, directly measure absorption phenomena, such as those associated with carbon dioxide (CO_2) and other gases. The atmosphere is nearly opaque to EM energy in part of the mid-IR and all of the far-IR regions. In the microwave region, by contrast, most of this radiation moves through unimpeded, so radar waves reach the surface (although weather radars are able to detect clouds and precipitation because they are tuned to observe backscattered radiation from liquid and ice particles).

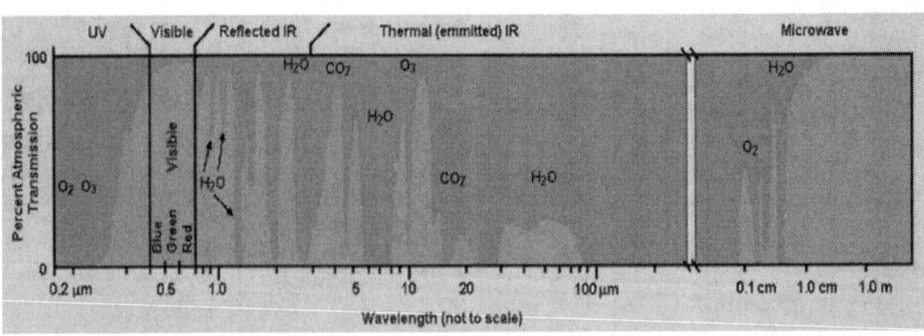

Figure 1.3 Atmospheric Windows. (See colour version on page 169)

Table 1.1 Major regions of the electromagnetic spectrum

Region Name	Wavelength	Comments
Gamma Ray	<0.03 nanometers	Entirely absorbed by the Earth's atmosphere and not available for remote sensing.
X-ray	0.03 to 30 nanometers	Entirely absorbed by the Earth's atmosphere and not available for remote sensing.
Ultraviolet	0.03 to 0.4 micrometers	Wavelengths from 0.03 to 0.3 micrometers absorbed by ozone in the Earth's atmosphere.
Photographic Ultraviolet	0.3 to 0.4 micrometers	Available for remote sensing the Earth. Can be imaged with cameras and sensors.
Visible	0.4 to 0.7 micrometers	Available for remote sensing the Earth. Can be imaged with cameras and sensors.
Near and Mid Infrared	0.7 to 3.0 micrometers	Available for remote sensing the Earth. Can be imaged with cameras and sensors.
Thermal Infrared	<0.7 to 3.0 micrometers	Available for remote sensing the Earth. This wavelength cannot be captured by film cameras. Sensors are used to image this wavelength band.
Micowave or Radar	0.1 to 100 centimeters	Longer wavelengths of this band can pass through clouds, fog, and rain. Images using this band can be made with sensors that actively emit microwaves.
Radio	>100 centimeters	Not normally used for remote sensing the Earth.

Traditional aerial photographs were black and white pictures based on camera and film technology. Such photographs related to one region of the EM spectrum. Satellite images are generally captured using sensors and digital technology. A sensor often records simultaneously in several different regions of the spectrum creating multi-images taken at the same time. The regions of the spectrum scanned are called "bands." A band is identified in nanometers (nm). For example, an image or band scanned in the visible light region of the spectrum would be identified as 400-700 nm. With multi-bands various color composite images can be created. Also, bands of various widths on the spectral scale can be recorded.

A single band image shows features in various grey tones but if several images are combined, they can form a color composite. Figure 1.4 B is a true color composite image of Charleston, South Carolina. A true color composite is based on using the red, green, and blue portions of the visible region of the EM spectrum. This type of composite relates to what the human eye would see if a person was on a satellite or aircraft looking down at the Earth. Other band combinations form false color composites. Figures 1.4 A, C, and D represent two different false color composites. A false color composite generally enhances certain features on an image, features that might not be as apparent on a true color composite. Using different color composites is one way that a remote sensing specialist detects features on the Earth.

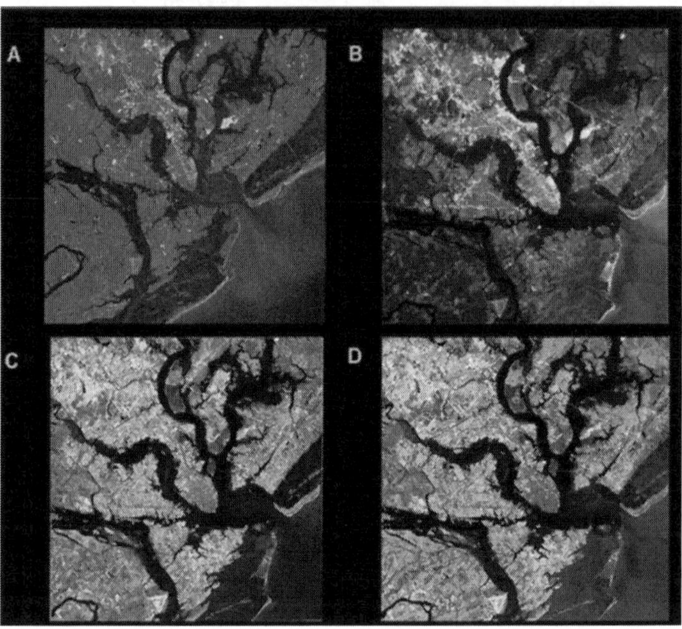

Figure 1.4 Color Composites. (See colour version on page 170)

Radiometric and Geometric Resolutions

Digital remote sensing deals with two types of resolutions: radiometric (spectral) and geometric (spatial). Radiometric resolution is the number of levels that a sensor can record spectral information. Such information generally ranges from 0-255 to 0-65,535. The most common range is 0-255, which relates to the storage capacity of an 8-bit computer byte. These numbers are integer values (whole numbers). A single byte can hold one distinct integer value ranging from 0-255. This value represents the degree of reflective or emitted energy recorded by a sensor for a particular ground spot on the Earth's surface. Although geographers take into consideration radiometric resolution when selecting imagery to study an environmental issue, they relate more toward geometric resolution due to its spatial nature.

Geometric resolution refers to the smallest amount of area on the Earth's surface for which a sensor can record radiometric (spectral) information. Generally this resolution is expressed in terms of a pixel (picture element). The pixel size of the Enhanced Thematic Mapper sensor on Landsat 7 is 30m, which relates to an area of 30m x 30m on the Earth's surface. In comparison, a sensor entitled AVHRR has a pixel of 1.1km^2 while the panchromatic sensor on the QuickBird satellite possesses a 61cm^2 pixel size.

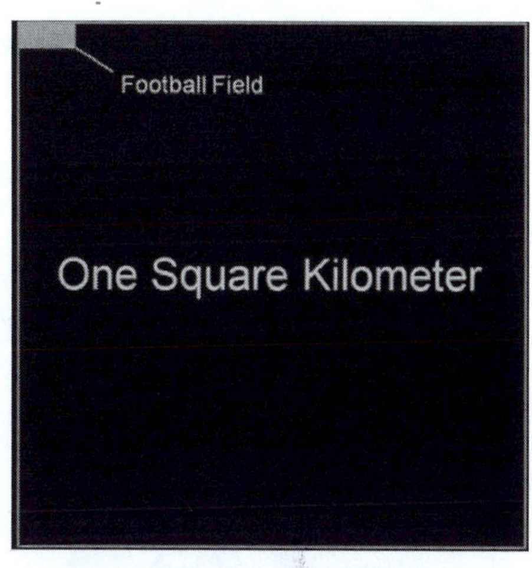

Figure 1.5 One Kilometer Pixel.

Geometric resolutions vary greatly and are defined loosely as being low, moderate, and high. The parameters associated with these designations change as finer resolutions in imagery become available. Imagery identified as being of moderate resolution at one point in time might now be low resolution. An image with a one kilometer pixel size is viewed as being a low resolution image. Such an image provides a synoptic coverage of the Earth's surface. which shows a massive dust storm blowing off the northwest African desert and blanketing hundreds of thousands of square miles of the eastern Atlantic Ocean with a dense dust.

Pixel Size

The given below figure is a representation of an American football field that has superimposed on it several different pixel sizes. The Landsat 7 ETM+ sensor records images at a 30m resolution, which is a moderate resolution size. It would take 5.5 Landsat 7 pixels to cover the area of a football field. One would not be able to differentiate much about the field with only 5.5 pixels. The four images shown in Figure 4 are Landsat images. They provide more detailed information than the low resolution image of the desert dust storm in Figure 1.6 but do not cover as much area. Situated in the upper left portion of the field is a QuickBird pixel. An image with a 60cm resolution would be a high resolution image. It would take 8,361 such pixels to cover a football field. With this number of pixels one would be able to see a great amount of detail about the field. Figure 1.7is a high resolution image of an Madurai Kamaraj

University. This image was acquired by a Geoeye image and the pixel resolution is 0.6m.

Figure 1.6 African Dust Storm (Low Resolution).
(See colour version on page 170)

Remote sensing imagery has many applications in cartography, land use and cover, agriculture, soils mapping, forestry, city planning, grassland management, archaeology, military observations, meteorology, and geomorphology, among other uses. In order to use such imagery, one must have considerable knowledge about the Earth's surface and a strong background in remote sensing data acquisition and analysis techniques. Since the Earth's surface consists of a mosaic of environmental conditions, geographers through their unique training in both the sciences and social sciences are well qualified to undertake various remote sensing applications.

Fundamental Considerations

Energy Source

Sensors can be divided into two broad groups—*passive* and *active*. Passive sensors measure ambient levels of existing sources of energy, while active ones provide their own source of energy. The majority of remote sensing is done with passive sensors, for which the sun is the major energy source. The earliest example of this is photography. With airborne cameras we have long been able to measure and record the reflection of light off earth features. While aerial photography is still a major form of remote sensing, newer solid state technologies have extended capabilities for viewing in the visible and near-infrared wavelengths to include longer wavelength solar radiation as well. However, not all passive sensors use energy from the sun. Thermal infrared and passive microwave sensors both measure natural earth energy emissions. Thus the passive sensors are simply those that do not themselves supply the energy being detected.

By contrast, active sensors provide their own source of energy. The most familiar form of this is flash photography. However, in environmental and mapping applications, the best example is RADAR. RADAR systems emit energy in the microwave region of the electromagnetic spectrum (Figure 1.2). The reflection of that energy by earth surface materials is then measured to produce an image of the area sensed.

Figure 1.7 Madurai Kamaraj University (High Resolution).
(See colour version on page 170)

Wavelength

As indicated, most remote sensing devices make use of electromagnetic energy. However, the electromagnetic spectrum is very broad and not all wavelengths are equally effective for remote sensing purposes. Furthermore, not all have significant interactions with earth surface materials of interest to us. Figure 1.2 illustrates the electromagnetic spectrum. The atmosphere itself causes significant absorption and/or scattering of the very shortest wavelengths. In addition, the glass lenses of many sensors also cause significant absorption of shorter wavelengths such as the ultraviolet (UV). As a result, the first significant *window* (i.e., a region in which energy can significantly pass through the atmosphere) opens up in the visible wavelengths. Even here, the blue wavelengths undergo substantial attenuation by atmospheric scattering, and are thus often left out in remotely sensed images. However, the green, red and near-infrared (IR) wavelengths all provide good opportunities for gauging earth surface interactions without significant interference by the atmosphere. In addition, these regions provide important clues to the nature of many earth surface materials. Chlorophyll, for example, is a very strong absorber of red visible wavelengths, while the near-infrared wavelengths provide important clues to the structures of plant leaves. As a result, the bulk of remotely sensed images used in GIS-related applications are taken in these regions.

Extending into the middle and thermal infrared regions, a variety of good windows can be found. The longer of the middle infrared wavelengths have proven to be useful in a number of geological applications. The thermal regions have proven to be very useful for monitoring not only the obvious cases of the spatial distribution of heat from industrial activity, but a broad set of applications ranging from fire monitoring to animal distribution studies to soil moisture conditions.

After the thermal IR, the next area of major significance in environmental remote sensing is in the microwave region. A number of important windows exist in this region and are of particular importance for the use of active radar imaging. The texture of earth surface materials causes significant interactions with several of the microwave wavelength regions. This can thus be used as a supplement to information gained in other wavelengths, and also offers the significant advantage of being usable at night (because as an active system it is independent of solar radiation) and in regions of persistent cloud cover (since radar wavelengths are not significantly affected by clouds).

Interaction Mechanisms

When electromagnetic energy strikes a material, three types of interaction can follow: reflection, absorption and/or transmission (Figure 1.8). Our main concern is with the reflected portion since it is usually this which is returned to the sensor system. Exactly how much is reflected will vary and will depend upon the nature of the material and where in the electromagnetic spectrum our measurement is being taken. As a result, if we look at the nature of this reflected component over a range of wavelengths, we can characterize the result as a *spectral response pattern.*

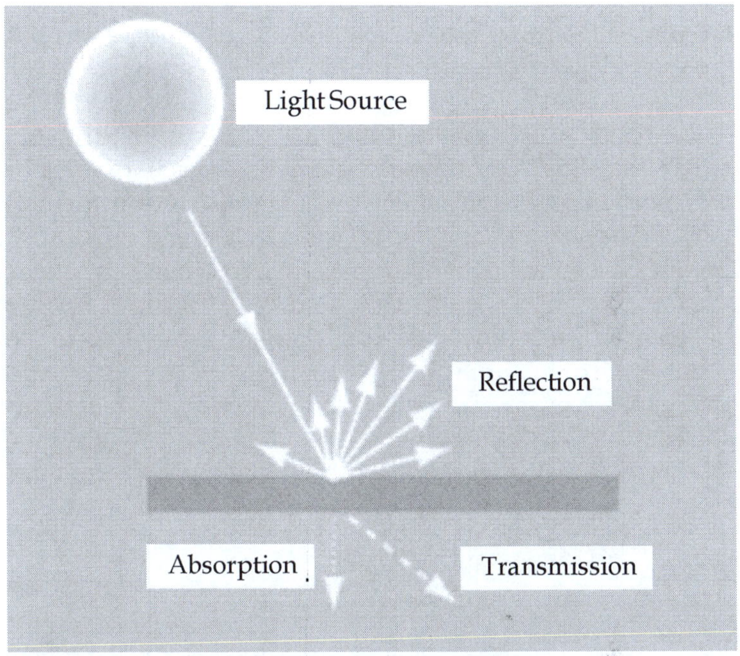

Figure 1.8 Interaction Mechanisms

Spectral Response Patterns

A spectral response pattern is sometimes called a *signature*. It is a description (often in the form of a graph) of the degree to which energy is reflected in different regions of the spectrum. Most humans are very familiar with spectral response patterns since they are equivalent to the human concept of color. For example, Figure 1.9 shows idealized spectral response patterns for several familiar colors in the visible portion of the electromagnetic spectrum, as well as for white and dark grey. The bright red reflectance pattern, for example, might be that produced by a piece of paper printed with a red ink. Here, the ink is designed to alter the white light that shines upon it and absorb the blue and green wavelengths. What is left, then, are the red wavelengths which reflect off the surface of the paper back to the sensing system (the eye). The high return of red wavelengths indicates a bright red, whereas the low return of green wavelengths in the second example suggests that it will appear quite dark.

The eye is able to sense spectral response patterns because it is truly a multi-spectral sensor (i.e., it senses in more than one place in the spectrum). Although the actual functioning of the eye is quite complex, it does in fact have three separate types of detectors that can usefully be thought of as responding to the red, green and blue wavelength regions. These are the *additive primary colors*, and the eye responds to mixtures of these three to yield a sensation of other hues. For example, the color perceived by the third spectral response pattern in Figure 1.9 would be a yellow—the result of *mixing* a red and green. However, it is important to recognize that this is simply our phenomenological perception of a spectral response pattern. Consider, for example, the fourth curve. Here we have reflectance in both the blue and red regions of the visible spectrum. This is a bimodal distribution, and thus technically not a specific hue in the spectrum. However, we would perceive this to be a purple! Purple (a color between violet and red) does not exist in nature (i.e., as a hue—a distinctive dominant wavelength). It is very real in our perception, however. Purple is simply our perception of a bimodal pattern involving a non-adjacent pair of primary hues.

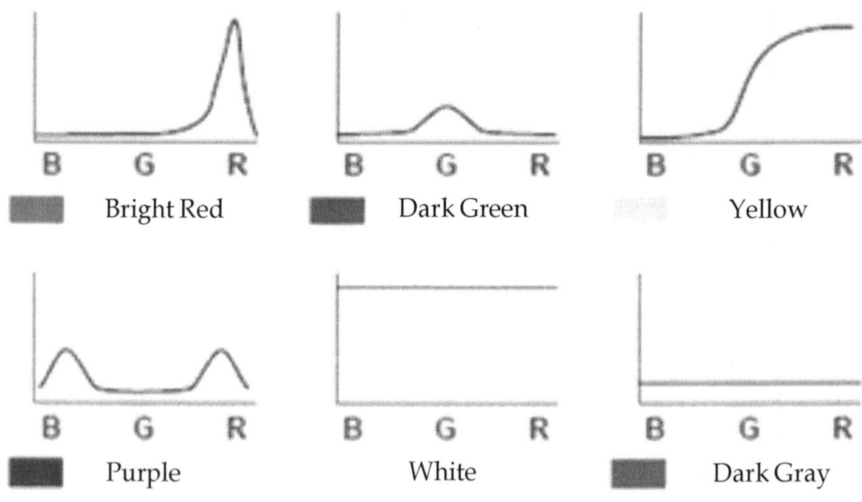

Figure 1.9 Spectral Response Patterns

In the early days of remote sensing, it was believed (more correctly *hoped*) that each earth surface material would have a distinctive spectral response pattern that would allow it to be reliably detected by visual or digital means. However, as our common experience with color would suggest, in reality this is often not the case. For example, two species of trees may have quite a different coloration at one time of the year and quite a similar one at another.

Finding distinctive spectral response patterns is the key to most procedures for computer-assisted interpretation of remotely sensed imagery. This task is rarely trivial. Rather, the analyst must find the combination of spectral bands and the time of year at which distinctive patterns can be found for each of the information classes of interest.

For example, Figure 1.10 shows an idealized spectral response pattern for vegetation along with those of water and dry bare soil. The strong absorption by leaf pigments (particularly chlorophyll for purposes of photosynthesis) in the blue and red regions of the visible portion of the spectrum leads to the characteristic green appearance of healthy vegetation. However, while this signature is distinctively different from most non-vegetated surfaces, it is not very capable of distinguishing between species of vegetation—most will have a similar colour of green at full maturation. In the near-infrared, however, we find a much higher return from vegetated surfaces because of scattering within the fleshy mesophyllic layer of the leaves.

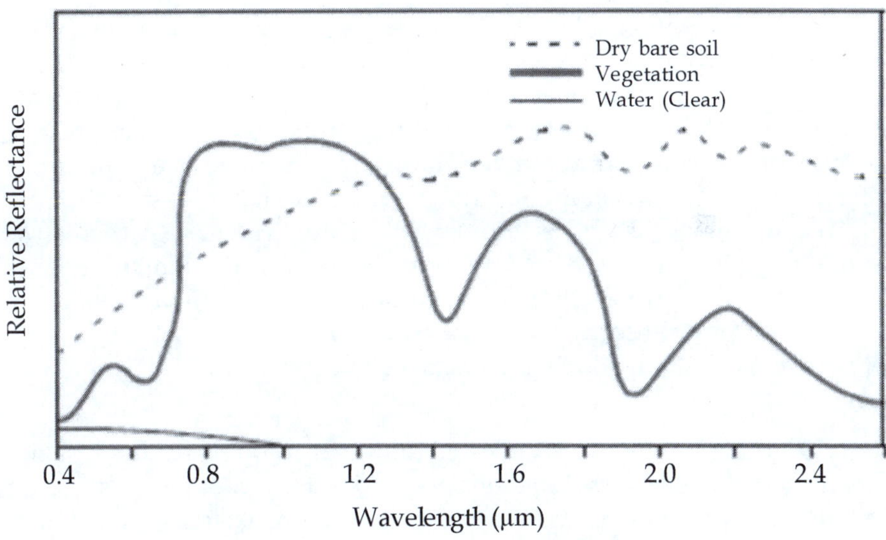

Figure 1.10 Spectral Signatures

Plant pigments do not absorb energy in this region, and thus the scattering, combined with the multiplying effect of a full canopy of leaves, leads to high reflectance in this region of the spectrum. However, the extent of this reflectance will depend highly on the internal structure of leaves (e.g., broadleaf versus needle). As a result, significant differences between species can often be detected in this region. Similarly, moving into the middle infrared region we see a significant dip in the spectral response pattern that is associated with leaf moisture.

This is, again, an area where significant differences can arise between mature species. Applications looking for optimal differentiation between species, therefore, will typically involve both the near and middle infrared regions and will use imagery taken well into the development cycle.

Multispectral Remote Sensing

In the visual interpretation of remotely sensed images, a variety of image characteristics are brought into consideration: color (or tone in the case of panchromatic images), texture, size, shape, pattern, context, and the like. However, with computer- assisted interpretation, it is most often simply colour (i.e., the spectral response pattern) that is used. It is for this reason that a strong emphasis is placed on the use of multispectral sensors (sensors that, like the eye, look at more than one place in the spectrum and thus are able to gauge spectral response patterns), and the number and specific placement of these spectral bands.

The LANDSAT satellite is a commercial system providing multi-spectral imagery in seven spectral bands at a 30 meter resolution. It can be shown through analytical techniques such as Principal Components Analysis, that in many environments, the bands that carry the greatest amount of information about the natural environment are the near-infrared and red wavelength bands. Water is strongly absorbed by infrared wavelengths and is thus highly distinctive in that region. In addition, plant species typically show their greatest differentiation here. The red area is also very important because it is the primary region in which chlorophyll absorbs energy for photosynthesis. Thus it is this band which can most readily distinguish between vegetated and non-vegetated surfaces.

Given this importance of the red and near-infrared bands, it is not surprising that sensor systems designed for earth resource monitoring will invariably include these in any particular multispectral system. Other bands will depend upon the range of applications envisioned. Many include the green visible band since it can be used, along with the other two, to produce a traditional false colour composite a full colour image derived from the green, red, and infrared ands (as opposed to the blue, green, and red bands of natural colour images). This format became common with the advent of colour infrared photography, and is familiar to many specialists in the remote sensing field. In addition, the combination of these three bands works well in the interpretation of the cultural landscape as well as natural and vegetated surfaces. However, it is increasingly common to include other bands that are more specifically targeted to the differentiation of surface materials.

For example, LANDSAT TM Band 5 is placed between two water absorption bands and has thus proven very useful in determining soil and leaf moisture differences. Similarly, LANDSAT TM Band 7 targets the detection of hydrothermal alteration zones in bare rock surfaces. By contrast, the AVHRR system on the NOAA series satellites includes several thermal channels for the sensing of cloud temperature characteristics.

Hyperspectral Remote Sensing

In addition to traditional multispectral imagery, some new and experimental systems such as AVIRIS and MODIS are capable of capturing *hyperspectral* data. These systems cover a similar wavelength range to multispectral systems, but in much narrower bands. This dramatically increases the number of bands (and thus precision) available for image classification (typically tens and even hundreds of very narrow bands). Moreover, hyperspectral signature libraries have been created in lab conditions and contain hundreds of signatures for different types of land covers, including many minerals and other earth materials. Thus, it should be possible to match signatures to surface materials with great precision. However, environmental conditions and natural variations in materials (which make them different from standard library materials) make this difficult. In addition, classification procedures have not been developed for hyperspectral data to the degree they have been for multispectral imagery. As a consequence, multispectral imagery still represents the major tool of remote sensing today.

Sensor/Platform Systems

Given recent developments in sensors, a variety of platforms are now available for the capture of remotely sensed data. Here we review some of the major sensor/platform combinations that are typically available to the GIS user community.

Wave Phenomena Concepts

Electromagnetic waves are *radiated* through space. When the energy encounters an object, even a very tiny one like a molecule of air, one of three reactions occurs. The radiation will either be reflected off the object, absorbed by the object, of transmitted through the object. The total amount of radiation that strikes an object is referred to as the *incident radiation*, and is equal to:

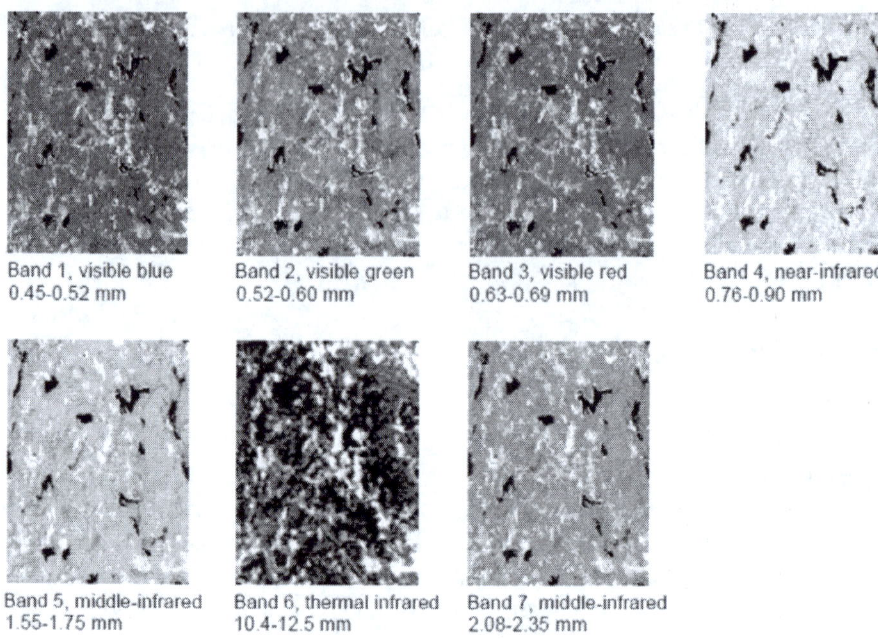

Figure 1.11 Band Combination

Reflected radiation + absorbed radiation + transmitted radiation

In remote sensing, we are largely concerned with REFLECTED RADIATION. This is the radiation that causes our eyes to see colors, causes infrared film to record vegetation, and allows radar images of the earth to be created.

The electric field and the magnetic field are important concepts that can be used to mathematically describe the physical effects of electromagnetic waves.

The electric field vibrates in a direction transverse (i.e. perpendicular) to the direction of travel of the electromagnetic wave.

The magnetic field vibrates in a direction transverse to the direction of the em wave AND transverse to the electric field.

POLARIZATION: Polarization is defined by the orientation of the electrical field E. It is usually described in terms of HORIZONTAL POLARIZATION and VERTICAL POLARIZATION. Polarization is most important when discussing RADAR applications of remote sensing.

Table 1.2. Frequencies and wavelength

Electromagnetic Waves, cont.

Some Frequencies and Wavelengths of Common E.M. Waves:

Type	Frequency	Wavelength
power lines	60 Hz	5×10^{6} m (about 3100 miles)
television	1 MHz	300 m
radar	1 GHz	0.3 m
infrared	3×10^{13} Hz	10^{-5} m
visible light	5×10^{14} Hz	0.6 µm (6000 Å)
ultraviolet	10^{15} Hz	0.3 µm (3000 Å)
gamma rays	3×10^{20} Hz	0.01 Å

Development of Remote Sensing in India

The pressure of growing population, increased demand for food, fodder and fuelwood combined with rapid industrialization and urbanization have caused tremendous pressure on natural resources and environment. The consequent impact of these activities have given rise to crisis in agro-ecosystems like land degradation, accelerated erosion, shifting cultivation, degradation of forests, increase in frequency of floods, pollution, urban sprawl etc. The challenge before us, therefore, is to reverse the process of degradation and undesirable changes in the environment in order to conserve and sustain the resource base of the country to meet the present demand as well as the future needs of the people.

In this context, the direct benefits from space technology and the applications driven approach adopted by the Indian Space Remote Sensing Programme have firmly established its immense potential to cater to the needs of national development. The evolution of National Natural Resource Management System (NNRMS) towards fully harnessing the potentials of space remote sensing and the development of the series of Indian Remote Sensing Satellites, besides establishment of necessary ground based data establishment of necessary ground based data

Amplitude and Wavelength

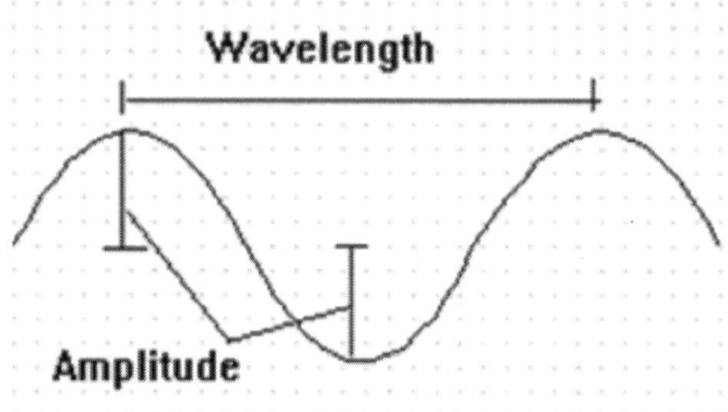

Figure 1.12 wavelength

Wave Descriptions

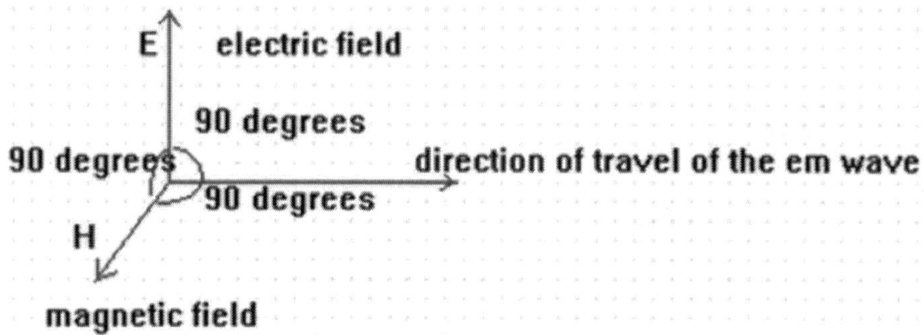

Figure 1.13 Wave Descriptions

reception processing and dissemination systems as well as remote sensing facilities at National Remote Sensing Agency (NRSA),

Space Applications Centre (SAC) and Regional Remote Sensing (Service Centres (RRSSCs) for efficient and effective analysis of remotely sensed data are the major steps accomplished in pursuit of this goal. With the establishment of Remote Sensing Applications Centres in several States under many Governmental organizations, remote sensing today has come to stay as an integral part of the national development efforts in the vital sectors of agriculture, hydrology, geology, forestry, oceanography, mineral resources and disasters management like drought, flood, cyclone, earthquake, landslides crop pests, forest fires etc., thus touching every facet of national development.

India has acquired a strong self reliant base to harness the full potential of this technology and as a result, the national objective of achieving sustainable development at micro level is being addressed through the integration of remotely sensed data with other relevant collateral information to arrive at locale specific, environment friendly, economically viable and culturally acceptable treatment packages (Rao, 1991 and Jayaraman, et al. 1993).

Space and Ground Segments

The Indian Remote Sensing Satellite (IRS-A), the first in the operational series of Indian Remote Sensing Satellites launched in March 1988, has been functioning satisfactorily for the past 6 ½ year but still continues tpo provide quality data. The second satellite in the IRS series, IRS-IB, launched in August 1991, is performing very well. These two satellites have become the mainstay of the National Natural Resources Management System (NNRMS) IRS-1A/1B provides imagery from the two cameras, linear Imaging Self Scanners, LISS-I with both the resolution 72.5m and LISS-IIA and LISS-IIB both with a resolution of 36.25m. IRS-1A and 1B together provide imagery with a combined repetivity of 11 days.

While IRS-IA has so far provided more than 5,00,000 scenes, IRS-1B has provided more than 2,00,000 scenes. The successful implementation of the operational IRS-IA 1B system in the country and the need to cater tot eh enhanced user demands in the 1990's has given rise to configuring the second generation remote sensing satellitesm IRS-IC and ID taking in to account the technology development scenario and user requirements during the nineties.

Future Satellite Programmes

The second generation of IRS Satellites namely IRs-1C and 1D which are now under development for launch in 1995 and 1998 respectively have made very good progress during the year. These satellites will have better spectral and spatial resolutions, more frequent revisits, stereo-viewing and onboard data recording facility. The imaging sensors planned for IRS-IC and ID are a multi-spectral Linear Imaging Self Scanner (LISS-3) in visible and near Ir bands, a short Wave infrared (SWIR) band, a panchromatic camera with stereo viewing capability and a Wide Field Sensor (WIFS) in visible and near IR(Jayaraman et. al 1992). It is also planned to have onboard data recording capabilities to record the pay load data for wider data coverage/distribution.

IRS-ID identical to IRS-IC will be in - orbit to ensure data continuity beyond the mid nineties. The development flights of PSLV during the timeframe of 1994-96 are progressing well. The IRS-P2 will carry the payload flown onboard IRS-IA and IB in order to provide continuity of data services to the remote sensing user community in the country. The IRS-P3 will carry a payload mix of sensors for application related to oceanography and vegetation dynamics, besides a scientific payload is planned to be included of experimental studies in X-ray astronomy. Thus IRS-P3 will carry a Modular Optoelectronic canner (MOS), a Wide field Sensor (WiFS) ad an X-ray astronomy payload.

Table 1.3 Orbital Characteristics of IRS-1C/1D

Orbit	:	Polar sun-synchronous
Altitude	:	817 km
Repetitivity	:	24 days
Local time of Equator Crossing	:	10.30 AM
Mission Life (planned)	:	3 years

Table 1.4 Characteristics of IRS-1C/1D Payloads

Characteristics	Payloads		
	Linear Imaging Self Scanning Sensor (LISS-3)	Wide Field Sensor (WiFS)	Panchromatic Camera (PAN)
No. of Spectral Bands	4	2	1
Spectral Bands (microns)	B2 0.52 - 0.59 B3 0.62 - 0.68 B4 0.77 - 0.86 B5 1.55 - 1.70	B3 0.62 - 0.68 B4 0.77 - 0.86	0.50 - 0.75
Spatial Resolution (m)	23.5 for B2, B3, B4, and 70.5 for B5	188	Better than 10m
Swath (Km)	142 km	774	70 km with steering Range +26°

The capability for microwave remote sensing has also been achieved in the India through development of airborne C-band SAR, multifrequency microwave radiometer and multifrequency scatterometer. Based on the encouraging results obtained from several pilot scale studies on ocean applications and considering the serious interest in monitoring and forecasting of a sea state from meteorological point of view, efforts are underway to realize and indigenously developed Oceansat.

Data Reception, Processing and Dissemination

The National Remote Sensing Agency (NRSA), Hyderabad continued to acquire and archives data from Indian Remote Sensing Satellites, IRS-IA and IRS-IB. In addition, data from other contemporary satellites, viz, Landsat NOAA and ERS-1 SAR are also being received. Augmention of ground segment elements for IRS-1C/1D data reception and processing is nearing completion. It is proposed to received SeaWiFS data after the launch SEASTAR satellite.

Satellite data products are being disseminated to the users in the form of photographic films, paper prints, CCT's floppy and cartridge products. About 705 of the user demands in the country is met by IRs-1A/1B data.

As part of aerial remote sensing programme, about 759 hours of aerial flying were undertake during the year mainly for aeromagnetic survey, town and country planning, sea level rise, oil exploration and metropolitan developments.

Indian Remote Sensing System

The Indian Remote sensing , Following the successful demonstration flights of Bhaskhar and Bhaskara-2 satellites launched in 1979 and 1981, respectively, India began to develop the indigenous Indian Remote Sensing (IRS) satellite program to support the national economy in the areas of agriculture, water resources, forestry and ecology, geology, water sheds, marine fisheries and coastal management.

Towards this end, India established the National Natural Resources Management System (NNRMS) for which the Department of Space (DOS) is the nodal agency, providing operational remote sensing data services.[1] Data from the IRS satellites is received and disseminated by several countries all over the world. With the advent of high-resolution satellites new applications in the areas of urban sprawl, infrastructure planning and other large scale applications for mapping have been initiated.

The IRS system is the largest constellation of remote sensing satellites for civilian use in operation today in the world, with 11 operational satellites. All these are placed in polar Sun-synchronous orbit and provide data in a variety of spatial, spectral and temporal resolutions. Indian Remote Sensing Programme completed its 25 years of successful operations on March 17, 2013.

Table 1.5 IRS Launch logs

S.No.	Satellite	Date of Launch	Launch Vehicle	Status
1	IRS 1A	17, March 1988	Vostok, USSR	Mission Completed
2	IRS 1B	29-Aug-91	Vostok, USSR	Mission Completed
3	IRS P1 (also IE)	20-Sep-93	PSLV-D1	Crashed, due to launch failure of PSLV
4	IRS-P2	15-Oct-94	PSLV-D2	Mission Completed
5	IRS-1C	28-Dec-95	Molniya, Russia	Mission Completed
6	IRS-P3	21-Mar-96	PSLV-D3	Mission Completed
7	IRS 1D	29-Sep-97	PSLV-C1	Mission Completed
8	IRS-P4 (Oceansat-1)	27-May-99	PSLV-C2	Mission Completed
9	Technology Experiment Satellite (TES)	22-Oct-01	PSLV-C3	Mission Completed
10	IRS P6 (Resourcesat-1)	17-Oct-03	PSLV-C5	In Service
11	IRS P5 (Cartosat 1)	5-May-05	PSLV-C6	In Service
12	Cartosat 2 (IRS P7)	10-Jan-07	PSLV-C7	In Service
13	Cartosat 2A	28-Apr-08	PSLV-C9	In Service
14	IMS 1	28-Apr-08	PSLV-C9	In Service
15	Oceansat-2	23-Sep-09	PSLV-C14	In Service
16	Cartosat-2B	12-Jul-10	PSLV-C15	In Service
17	Resourcesat-2	20-Apr-11	PSLV-C16	In Service
18	Megha-Tropiques	12-Oct-11	PSLV-C18	In Service
19	RISAT-1	26-Apr-12	PSLV-C19	In Service
20	SARAL	25-Feb-13	PSLV-C20	In Service

INSAT

The **Indian National Satellite System**, or **INSAT**, is a series of multipurpose geo-stationary satellites launched by ISRO to satisfy the telecommunications, broadcasting, meteorology, and search and rescue operations. Commissioned in 1983, INSAT is the largest domestic communication system in the Asia Pacific Region. It is a joint venture of the Department of Space, Department of Telecommunications, India Meteorological Department, All India Radio and Doordarshan. The overall coordination and management of INSAT system rests with the Secretary-level INSAT Coordination Committee.

INSAT satellites provide transponders in various bands (C, S, Extended C and K_u) to serve the television and communication needs of India. Some of the satellites also have the Very High Resolution Radiometer (VHRR), CCD cameras for metrological imaging. The satellites also incorporate transponder(s) for receiving distress alert signals for search and rescue missions in the South Asian and Indian Ocean Region, as ISRO is a member of the Cospas-Sarsat programme.

INSAT system

The Indian National Satellite (INSAT) system was commissioned with the launch of INSAT-1B in August 1983 (INSAT-1A, the first satellite was launched in April 1982 but could not fulfil the mission). INSAT system ushered in a revolution in India's television and radio broadcasting, telecommunications and meteorological sectors. It enabled the rapid expansion of TV and modern telecommunication facilities to even the remote areas and off-shore islands. Together, the system provides transponders in C, Extended C and K_u bands for a variety of communication services. Some of the INSATs also carry instruments for meteorological observation and data relay for providing meteorological services. KALPANA-1 is an exclusive meteorological satellite. The satellites are monitored and controlled by Master Control Facilities that exist in Hassan and Bhopal.

LANDSAT

The **Landsat program** is the longest-running enterprise for acquisition of satellite imagery of Earth. On July 23, 1972 the Earth Resources Technology Satellite was launched. This was eventually renamed to Landsat. The most recent, Landsat 8, was launched on February 11, 2013. The instruments on the Landsat satellites have acquired millions of images. The images, archived in the United States and at Landsat receiving stations around the world, are a unique resource for global change research and applications in agriculture, cartography, geology, forestry, regional planning, surveillance and education, and can be viewed through the USGS 'EarthExplorer' website. Landsat 7 data has eight spectral bands with spatial resolutions ranging from 15 to 60 meters; the temporal resolution is 16 days.

Table 1.6 Satellite chronology

Instrument	Launched	Terminated	Duration	Notes
Landsat 1	23-Jul-72	6-Jan-78	2 years, 11 months and 15 days	Originally named Earth Resources Technology Satellite 1.
Landsat 2	22-Jan-75	25-Feb-82	2 years, 10 months and 17 days	Nearly identical copy of Landsat 1
Landsat 3	5-Mar-78	31-Mar-83	5 years and 26 days	Nearly identical copy of Landsat 1 and Landsat 2
Landsat 4	16-Jul-82	14-Dec-93	11 years, 4 months and 28 days	
Landsat 5	1-Mar-84	June 5, 2013[7]	29 years, 3 months and 4 days	Nearly identical copy of Landsat 4. Longest Earth-observing satellite mission in history.
Landsat 6	5-Oct-93	5-Oct-93	0 days	Failed to reach orbit.
Landsat 7	15-Apr-99	Still active	17 years, 2 months and 15 days	Operating with scan line corrector disabled since May 2003.[8]
Landsat 8	11-Feb-13	Still active	3 years, 4 months and 19 days	Originally named Landsat Data Continuity Mission from launch until May 30, 2013, when NASA operations were turned over to USGS.[9]

SPOT

SPOT (French: *Satellite Pour l'Observation de la Terre*, lit. "Satellite for observation of Earth") is a commercial high-resolution optical imaging Earth observation satellite system operating from space. It is run by Spot Image, based in Toulouse, France. It was initiated by the CNES (*Centre national d'études spatiales* – the French space agency) in the 1970s and was developed in association with the SSTC (Belgian scientific, technical and cultural services) and the Swedish National Space Board (SNSB). It has been designed to improve the knowledge and management of the Earth by exploring the Earth's resources, detecting and forecasting phenomena involving climatology and oceanography, and monitoring human activities and natural phenomena. The SPOT system includes a series of satellites and ground control resources for satellite control and programming, image production, and distribution. Earlier satellites were launched using the European Space Agency's Ariane 2, 3, and 4 rockets, while SPOT 6 and SPOT 7 were launched by the Indian PSLV.

SPOT Image is marketing the high-resolution images, which SPOT can take from every corner of the Earth.

- SPOT 1 launched February 22, 1986 with 10 panchromatic and 20 meter multispectral picture resolution capability. Withdrawn December 31, 1990.
- SPOT 2 launched January 22, 1990 and deorbited in July 2009.
- SPOT 3 launched September 26, 1993. Stopped functioning November 14, 1997.
- SPOT 4 launched March 24, 1998. Stopped functioning July, 2013.
- SPOT 5 launched May 4, 2002 with 2.5 m, 5 m and 10 m capability. Stopped functioning March 31, 2015.
- SPOT 6 launched September 9, 2012.
- SPOT 7 launched on June 30, 2014.

IKONOS

IKONOS is a commercial earth observation satellite, and was the first to collect publicly available high-resolution imagery at 1- and 4-meter resolution. It offers multispectral (MS) and panchromatic (PAN) imagery. The IKONOS launch was called "one of the most significant developments in the history of the space age". IKONOS imagery began being sold on January 1, 2000.

Specifications

Spacecraft

IKONOS is a three-axis stabilized spacecraft designed by Lockheed Martin. The design later became known as the LM900 satellite bus system. The satellite's altitude is measured by two star trackers and a sun sensor and controlled by four reaction wheels; location knowledge is provided by a GPS receiver. The design life is seven years; S/C body size=1.83 m × 1.57 m (hexagonal configuration); S/C mass = 817 kg; power = 1.5 kW provided by three solar panels. The LM900 spacecraft is a three-axis stabilized bus that is designed to carry scientific payloads in LEOs. It provides precision pointing on an ultra stable highly agile platform. Payloads for a variety of scientific and remote sensing applications may be accommodated including laser sensors, imagers, radar sensors, electro-optical and astronomical sensors, as well as planetary sensors. The LM900 bus shares a hardware heritage with Iridium, which is the basis for the LM700 bus.

Communications

IKONOS conducts telemetry, tracking and control in the 8345.968–8346.032 MHz band (downlink) and 2025–2110 MHz band (uplink). Downlink data carrier operates in the 8025-8345 MHz band.

Optics & Detectors

IKONOS has a primary mirror aperture of 0.7 m (2.3 feet), and a folded optical focal length of 10 m (about 33 feet) using 5 mirrors. The main mirror features a honeycomb design to reduce mass. The detectors at the focal plane include a pan-chromatic and a multi-spectral sensor, with 13500 pixels and 3375 pixels respectively (cross-track). Total instrument mass is 171 kg (377 pounds) and it uses 350 watts.

QuickBird

QuickBird was a high-resolution commercial earth observation satellite, owned by DigitalGlobe launched in 2001 and decayed in 2015. It was the first satellite in a constellation of three scheduled to be in orbit by 2008. QuickBird used Ball Aerospace's Global Imaging System 2000 (BGIS 2000). The satellite collected panchromatic (black and white) imagery at 61 centimeter resolution and multispectral imagery at 2.44- (at 450 km) to 1.63-meter (at 300 km) resolution, as orbit altitude is lowered during the end of mission life.

At this resolution, detail such as buildings and other infrastructure are easily visible. However, this resolution is insufficient for working with smaller objects such as a license plate on a car. The imagery can be imported into remote sensing image processing software, as well as into GIS packages for analysis. The imagery can also be used as a backdrop for mapping applications, such as Google Earth and Google Maps.

❑❑❑

2

Aerial Remote Sensing

Introduction

Although both maps and aerial photos present a "bird's-eye" view of the earth, aerial photographs are NOT maps. Maps are orthogonal representations of the earth's surface, meaning that they are directionally and geometrically accurate (at least within the limitations imposed by projecting a 3-dimensional object onto 2 dimensions). Aerial photos, on the other hand, display a high degree of radial distortion. That is, the topography is distorted, and until corrections are made for the distortion, measurements made from a photograph are not accurate. Nevertheless, aerial photographs are a powerful tool for studying the earth's environment.

Because most GISs can correct for radial distortion, aerial photographs are an excellent data source for many types of projects, especially those that require spatial data from the same location at periodic intervals over a length of time. Typical applications include land use surveys and habitat analysis.

This unit discusses benefits of aerial photography, applications, the different types of photography, and the integration of aerial photographs into GISs.

Basic Elements of Air Photo Interpretation

Novice photo interpreters often encounter difficulties when presented with their first aerial photograph. Aerial photographs are different from "regular" photos in at least three important ways:

Objects are portrayed from an overhead (and unfamiliar) position

- Very often, infrared wavelengths are recorded, and
- Photos are taken at scales most people are unaccustomed to seeing

These "basic elements" can aid in identifying objects on aerial photographs.

Tone (also called Hue or Color) – Tone refers to the relative brightness or color of elements on a photograph. It is, perhaps, the most basic of the interpretive elements because without tonal differences none of the other elements could be discerned.

Size – The size of objects must be considered in the context of the scale of a photograph. The scale will help you determine if an object is a stock pond or Lake Minnetonka.

Shape – Refers to the general outline of objects. Regular geometric shapes are usually indicators of human presence and use. Some objects can be identified almost solely on the basis of their shapes.

- The Pentagon Building
- (American) football fields
- Cloverleaf highway interchanges

Texture – The impression of "smoothness" or "roughness" of image features is caused by the frequency of change of tone in photographs. It is produced by a set of features too small to identify individually. Grass, cement, and water generally appear "smooth", while a forest canopy may appear "rough".

Pattern *(spatial arrangement)* – The patterns formed by objects in a photo can be diagnostic. Consider the difference between (1) the random pattern formed by an unmanaged area of trees and (2) the evenly spaced rows formed by an orchard.

Shadow – Shadows aid interpreters in determining the height of objects in aerial photographs. However, they also obscure objects lying within them.

Site – Refers to topographic or geographic location. This characteristic of photographs is especially important in identifying vegetation types and landforms. For example, large circular depressions in the ground are readily identified as sinkholes in central Florida, where the bedrock consists of limestone. This identification would make little sense, however, if the site were underlain by granite.

Association – Some objects are always found *in* association with other objects. The context of an object can provide insight into what it is. For instance, a nuclear power plant is not (generally) going to be found in the midst of single-family housing.

Advantages of Aerial Photography over Ground-Based Observation

- Aerial photography offers an improved vantage point.
- Aerial photography has the capability to stop action.
- It provides a permanent recording.
- It has broader spectral sensitivity than the human eye.
- It has better spatial resolution and geometric fidelity than many ground-based sensing methods.

Types of Aerial Photography

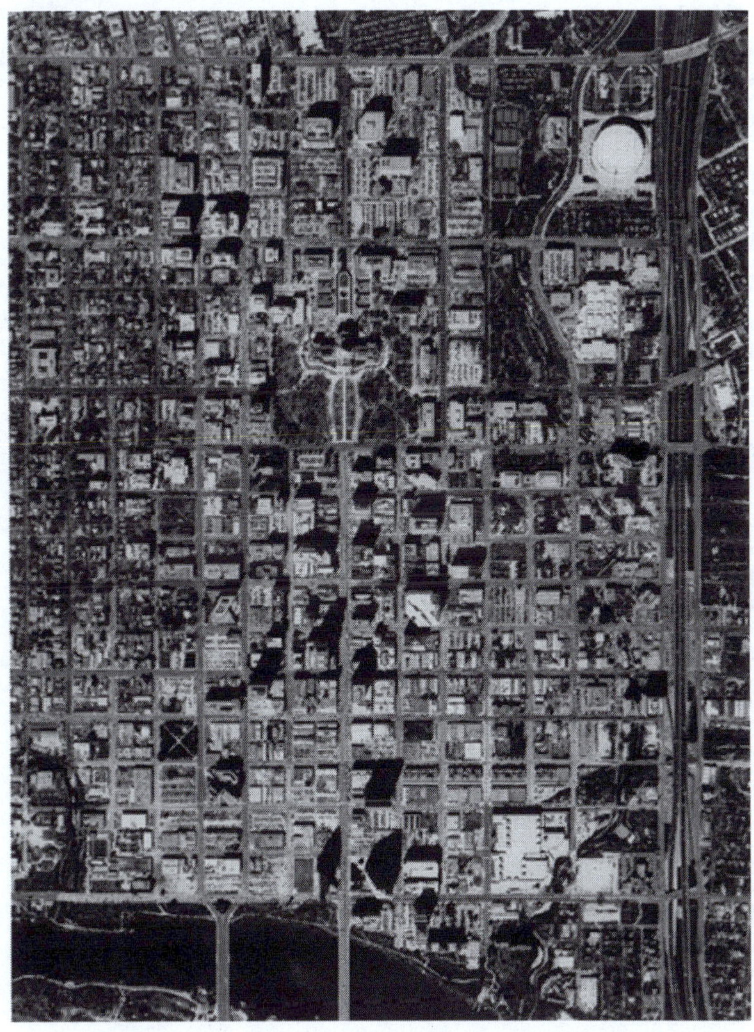

Figure 2.1 Black and White Photos Austin, Texas

Color Infrared

In 1903 or 1904 the first reliable black and white infrared film was developed in Germany. The film emulsion was adjusted slightly from regular film to be sensitive to wavelengths of energy just slightly longer than red light and just beyond the range of the human eye. By the 1930s, black and white IR films were being used for landform studies, and from 1930 to 1932 the National Geographic Society sponsored a series of IR photographs taken from hot air balloons.

Throughout the 1930s and 1940s, the military was hard at work developing color infrared film, eager to exploit it for surveillance. By the early 1940s the military was successful in its attempts. It developed a film that was able to distinguish camouflaged equipment from surrounding vegetation. Within months, however, an IR reflecting paint was developed for use on military vehicles, effectively making IR film technology useless to the military. So, they dropped it.

The scientific community, however, has made continuous use of the film technology.

Color infrared film is often called "false-color" film. Objects that are normally red appear green, green objects (except vegetation) appear blue, and "infrared" objects, which normally are not seen at all, appear red.

The primary use of color infrared photography is vegetation studies. This is because healthy green vegetation is a very strong reflector of infrared radiation and appears bright red on color infrared photographs.

Basic Photogrammetry

Applications of Aerial Photography

Introduction: The Scope of Air Photography

- Land-Use Planning and Mapping
- Geologic Mapping
- Archaeology
- Species Habitat Mapping
- Integration of Aerial Photography into GIS

Flight lines and aerial photograph coverage

Most aerial photography is obtained in order to produce topographic maps. This so called 'mapping photography' is best accomplished with a short focal-length camera lens (6 inches or shorter) and at a flight height that gives photographs of appropriate scale for use with specific map-making instruments. The scale of this photography commonly is 1:20 000 or smaller and may be as small as 1:60 000. Mapping photography generally is taken in the spring or fall when obscuring leafy foliage is at a minimum.

Normally photographic flights are made in a north-south or east-west direction unless it is for a special-purpose project such as a floodplain map in a river valley. The flight path for taking aerial photographs consists of a series of parallel flight lines laid out so as to obtain overlapping photographs that can be used for stereoscopic viewing.

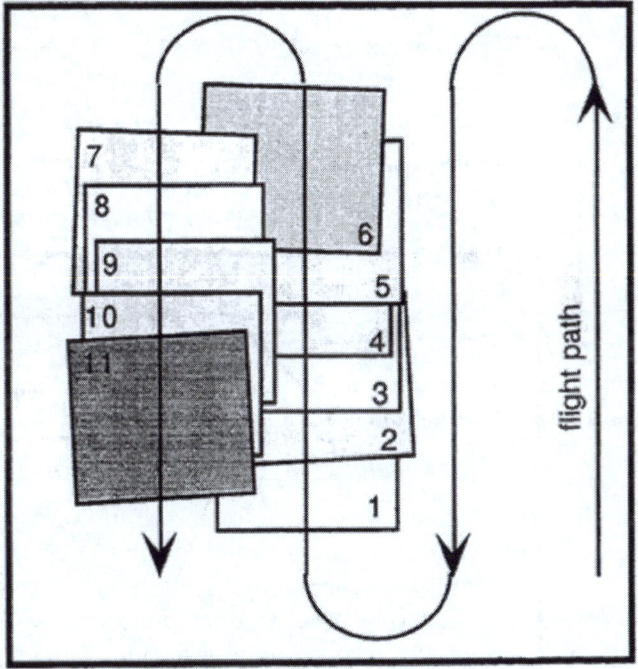

Figure 2.2 Flight path for an aerial photographic survey showing 60% overlap along the flight lines and 25-30% side lap between adjacent flight lines.

Ideally each flight line should be a straight course but inevitably cross-winds and other problems of aircraft navigation can modify the flight path. It is not unusual, for example, to find that there has been 'crab' or 'drift' of the aircraft during the photographic mission.

Crab occurs when the aircraft is slightly turned into a cross-wind in order to maintain a proper heading. If the airborne camera is not turned to compensate for such crabbing, the edges of the photographs will not parallel the flight path (Figure 2.2). Drift occurs when the aircraft is not turned to compensate for a cross-wind component and the plane is blown sideways and off the course of the flight line (Figure 2.3).

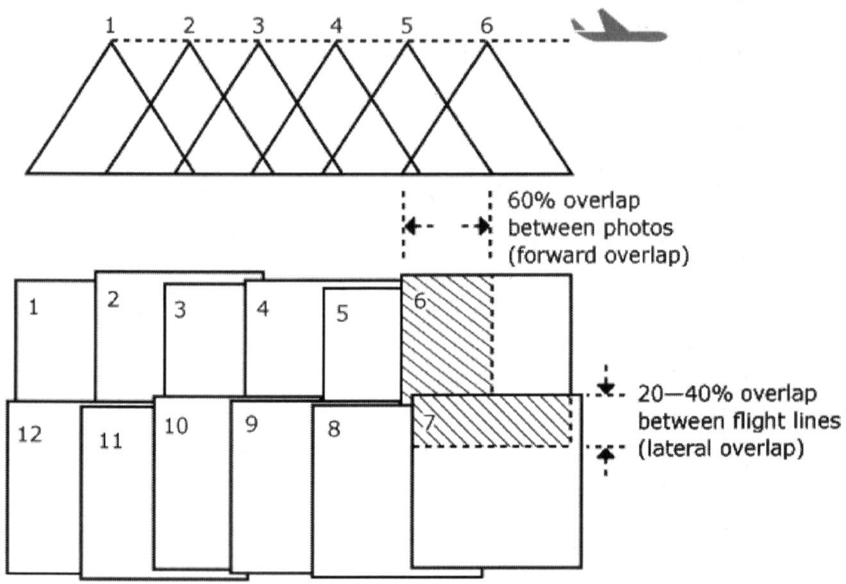

Figure 2.3 The relation of photograph orientation to the correct and actual flight lines of an aerial survey under cross-wind conditions causing crab and drift.

All Canadian Government producers of aerial photography inscribe each photograph with a serial number. Generally this consists of a letter followed by a five-digit number identifying the flight and the photograph number. Most recent B.C. Government aerial photography has a 'BC' serial number and each photograph displays marginal data including the date and time of photography, as well as images of a bubble level-indicator and an altimeter.

Geometry of the single aerial photograph

The geometry of a single vertical aerial photograph is illustrated in Figure 2.4. In a vertical aerial photograph the optical axis of the camera is vertical and the plane of the photograph (film) is horizontal. The point where the optical axis intersects the photograph is termed the *centre point* or *principal point* of the photograph. This can be located on an aerial photograph as the intersection of lines drawn between opposite *fiducial marks* in the margins of the print. In a perfectly vertical aerial photograph the principal point also represents the *plumb point* or *nadir point* which is the photographic position representing the point on the earth's surface vertically beneath the camera lens at the time of exposure. In practice a vertical aerial photograph is rarely absolutely vertical and the nadir point and the centre point do not coincide exactly, the usually small difference being the result of *tilt*. The distance between the camera lens and the ground represents the *flight height* of the aircraft and the *focal length* is the distance between the camera lens and the film.

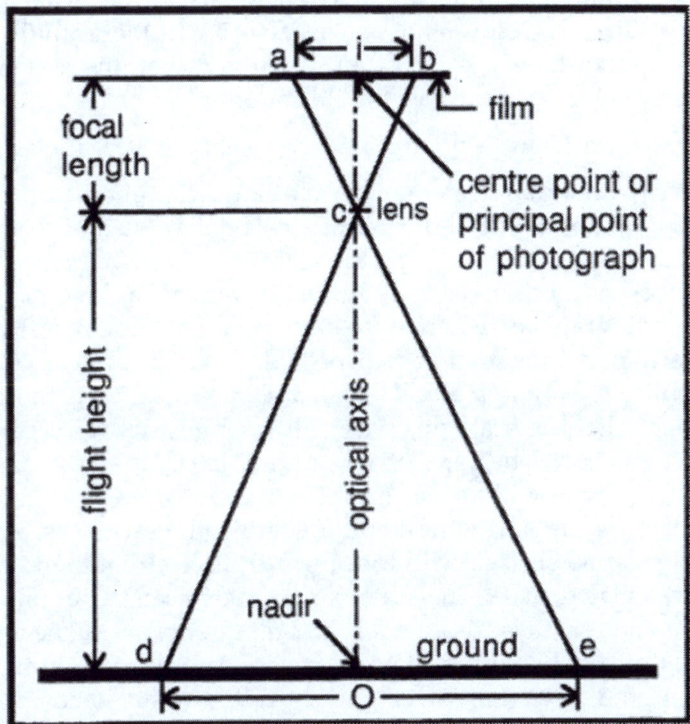

Figure 2.4 Geometry of a single vertical aerial photograph

The scale of a vertical aerial photograph

One of the most significant geometric relationships is that equal angles are subtended at a camera lens by an object and by its photographic image. In other words, the triangles abc and cde are similar and it follows that the ratio of object size (O) to image size (i) is the same as the ratio of focal length (f) to flight height (H), or f/H = I O

The ratio of image to object size is the general scale of the aerial photograph and it follows that the scale may be determined if the camera focal length and flight height are known:

scale = focal length/flight height

For example, the average scale of photographs taken at 10 000 feet (3 048 metres) above the terrain with a 6-inch (0.1524-metre) focal-length lens is 0.1524 / 3 048 or 1/ 20 000. Note that, for a photo-scale determination using equation, the focal length and the flight height must be expressed in the same units. Thus, if we had chosen to use the Imperial units, the scale would be 0.5 feet / 10 000 feet = 1: 20 000. It also is very important to remember that flight height refers to the distance above the ground directly below and not necessarily to the altitude (height above sea level or the base airport) of the aircraft. For this reason a more precise restatement of equation should read:

Scale = focal length / [flight altitude - average terrain elevation]

It is very important to remember that an aerial photograph is not a controlled map and that a photographic scale determined in this way is a *mean* or *averaged* scale . That is, it is the scale of the *mean surface* of the area photographed and not necessarily the accurate scale at any one point. If the area photographed is horizontal and very flat (low relief) then the scale determined by equation will provide a good indication of scale over the entire photograph. But if the area photographed has considerable topographic relief the scale may vary considerably above or below that given by equation. The tops of mountains that are above the mean topographic surface will appear on the aerial photograph at larger than the calculated scale and valley bottoms and other surfaces below the mean topographic surface will appear on the aerial photograph at less than the calculated scale. The scale variation over an aerial photograph of a mountainous area is considerable and represents one of the important limitations of aerial photographs as maps. An alternative method of determining the scale of an aerial photograph is by direct comparison of the photograph with a topographic map of known scale. This method commonly is used because information on the focal length and flight height of photography usually are not readily available.

Scale of a photograph is determined by this method as follows:

1. Select two points common to both the photograph and the topographic map. They should be as widely spaced as possible in order to keep measurement error to a minimum. Using the scale of the map measure the terrain distance, Δt, between the two points.

2. Measure the distance on the photograph, Δp, between the two points in question.

3. If Δt and Δp are *expressed in the same units*, the scale of the aerial photograph is € Δp Δt with the numerator reduced to unity in the conventional way. For example, if the distance between two points, A and B, is measured on an aerial photograph to be 9.26 cm, and a topographic map shows the corresponding distance on the ground to be 3.24 km, then

   Scale of the aerial photograph = 9.26/324 000 = 1/34 989 or

   Approximately 1/35 000

Photographic scale determined by the map-comparison method is that for the sites A to B on which it is based. For other sites on the same aerial photograph the scale may be smaller or larger depending on whether these sites are lower or higher in elevation with respect to A-B. The degree of variation in scale is directly proportional to the amount of relief in the area.

Measurement of heights on single aerial photographs

The scale variation on an aerial photograph caused by relief results from the horizontal displacement of images from their correct photograph position as depicted in Figure 2.5. The basic difference between an aerial photograph and a map in this regard can be demonstrated by comparing the central projection of the single photograph, in which all objects are positioned as though viewed from the same point, with the orthographic projection of a map, in which all objects on the ground are positioned as though viewed from vertically above. In a vertical aerial photograph the displacement of images is in a radial direction from the centre point of the photograph. This displacement is termed the *radial displacement due to relief* and represents an error in map positioning. For example, point A on the ground depicted in Figure 2.5 has a true vertically-projected horizontal position on the mean topographic surface indicated by A′ and a corresponding true image location on the photograph indicated by a′. But point A is above the mean topographic surface and its position on this datum plane appears to be A″ and is indicated on the photograph by an image at a″. In other words, the terrain feature A on the ground appears to be displaced radially *outwards* from its true map position (A′) to an apparent and incorrect position at A″. On the photograph this radial displacement due to relief appears as the interval a′ to a″. Similarly, point B which falls below the datum plane of the photograph, has a true location B′ on the mean topographic surface and should appear as b′ on the photograph. Instead, it has an apparent location at B″ and appears on the photograph at b″, this time the result of *inward* radial displacement due to relief. On any one aerial photograph the degree of displacement due to relief increases with increasing distance from the centre point and with increasing difference in elevation above or below the datum. For example, points A and B in Figure 2.5 B are the same height above the datum plane but point A is more distant from the ground nadir than B. Thus the radial displacement due to relief is greater for point A than for B. Note also that C is the same distance from the ground nadir as A but because C is not as high

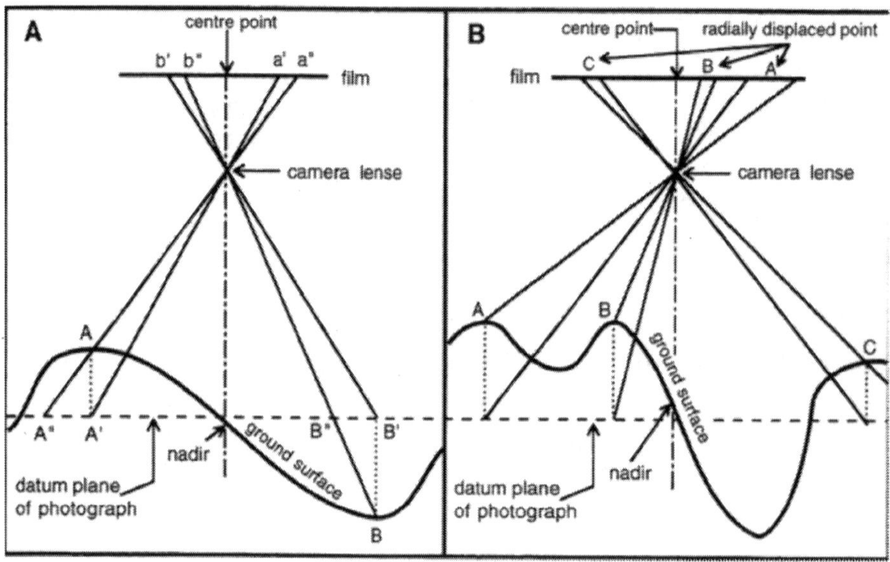

Figure 2.5 Radial displacement

Radial displacement on aerial photographs. A: Relative positions of correctly plotted features in the orthographic projection of a map and the corresponding displaced images on an aerial photograph. **B**: Relative radial displacement of terrain features on an aerial photograph in relation to height above the datum plane and distance from the nadir.

Above the datum plane as A, it is not radially displaced to the same extent. For any one aerial photograph the amount of radial displacement, m, of the top of an object from its base, can be determined by the relation:

$$m = \frac{r\,h}{H}$$

In which r = radial distance on the photograph from the centre point to the top of the image displaced, h = height of the object displaced, and

H = the flight height. Rearranging equation yields a convenient expression for estimating the height of an object on a photograph by measuring its radial displacement:

$$h = \frac{m}{r} H$$

In equations the photographic measurements m and r must be expressed in the same units and h will be in the units of H. An example of this type of height determination appears in Figure 2.6. Clearly, this method of estimating object height depends in part on an accurate determination of the height of the camera above the object photographed. This local height is most accurately determined by finding the scale of the photograph by the map-comparison method and solving equation for the known focal length of the camera. It is also clear that this method of determining the height of an object must be based on measurable displacement; the object top must in fact be vertically above the base and both must be visible in order to measure the displacement. Ideal objects are such things as trees, vertical cliffs, buildings, bridges, flag poles, and Hydro towers.

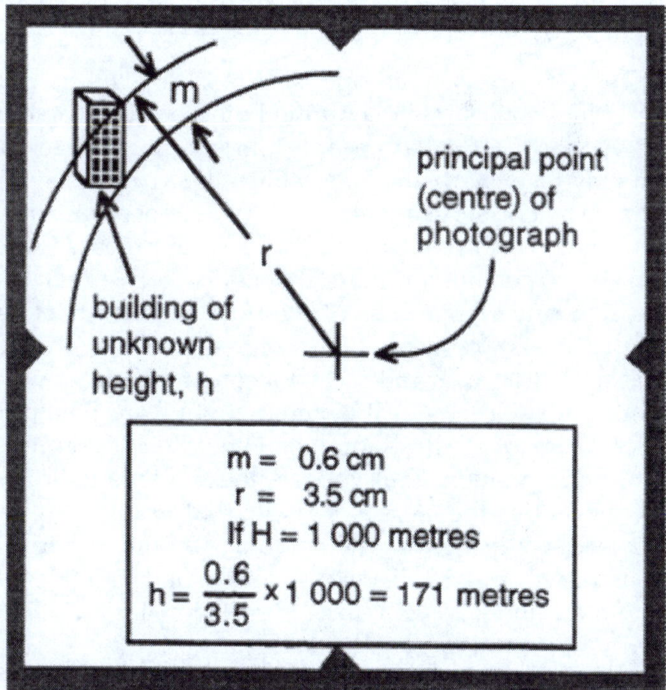

Figure 2.6 Radial displacement geometry used to determine object height on a single aerial photograph

Alternative means of determining the height of an object on a single vertical aerial photograph is by measuring shadow lengths. On any one, just as we use parallax displacement as a cue to perceiving depth in ordinary vision (the same view through two spatially separated eyes), a *stereoscope* reproduces the same parallax dependent sense of depth by allowing us to view one photograph with the left eye while viewing another spatially displaced photograph with the right eye. This technique successfully tricks the brain into thinking that the parallax displacement apparent in the merged images of the two adjacent photographs is the result of viewing a truly three-dimensional surface through unaided eyes. Consequently, stereoscopic viewing allows us to see the area photographed in three dimensions; it is a very powerful analytical technique that is an important part of the geographer's 'tools of trade'.

The geometry of the stereoscopic model allows us to determine the height of the land surface based on parallax displacement between adjacent photographs. This principle is the basis of all photogrammetry and topographic mapping from aerial photographs. Although the application of these photogrammetric techniques are beyond the scope of our introductory survey of aerial photography (they are developed in Geography 350), we will be concerned with stereoscopic viewing as an aid to aerial photograph interpretation.

The purpose of the stereoscope is to facilitate the merging of the two photographic images into one unified view. Stereoscopic vision can be achieved by most (but not all) people after a little perseverance. Initial stereovision may take some time to establish but once achieved it can be readily reestablished on subsequent occasions. Stereoscopes may be simple but quite effective pocket stereoscopes for viewing prepared stereogram's or they may be larger mirror stereoscopes for convenient viewing of pairs of contact prints. Regardless of the type of instrument being used the procedure for stereoscopic viewing is the same. The pair of photographs should be arranged so that a common field of vision can be observed through each lens of the stereoscope. It may help to place a finger on each image of a feature common to both photographs and adjust the photograph orientation until you see the fingernails on each finger merge into one under the stereoscope. At that point the photographs should be correctly adjusted for stereoscopic viewing. There will be no mistaking the onset of stereovision; the entire field of stereovision will be thrown into an unequivocally three dimensional view.

You should note that the heights of objects and the relief of the land are exaggerated under the stereoscope. This may be a virtue in photographs of low relief terrain but it must always be considered when interpreting photographs of any area.

Aerial photograph identification and interpretation

Rarely do we identify and interpret the information on aerial photographs without some clear purpose in mind. A physical geographer, for example, may be interested in mapping vegetation types or changes in river boundaries or the occurrence of glaciated terrain. A geographer concerned with settlement may be interested in measuring housing density or crop areas or the encroachment of urban areas into farmer agricultural lands. The reasons are many and varied and the appropriate requisite skills of aerial photograph interpretation are similarly diverse.

The best teacher of recognition and interpretive skills in any of these contexts is experience and later exercises involve some 'hands on' activity in this regard. Nevertheless, there are some useful general principles of interpretation common to any work with aerial photograph interpretation and you should note them well.

Photographs should always be arranged so that any shadows fall towards the viewer; like hill shading, shadows can help to convey a sense of the third dimension. Indeed, on some photographs the relief may appear to be inverted if the shadows are falling away from the viewer!

Just as we use the familiar to evaluate the unknown in all other areas of human activity, we should not disregard the obvious or apparently mundane information when we interpret aerial photographs. Familiar features such as roads and houses provide useful scales. Many other cultural features can be identified by association. For example, noting the presence of an adjacent railroad siding or a playground may be a means of determining whether a building is a school or a factory. The ability to pay attention to detail is important and is the single most important quality distinguishing the novice from the experienced interpreter of aerial photographs.

Tonal pattern on conventional aerial photography records reflectivity and not necessarily colour. The degree to which light is reflected from surfaces depends in turn on the roughness or texture of the surface. If a surface in very smooth, like an asphalt parking lot, it may reflect a great deal of light and will appear very light in tone even though it may be close to black in colour. By the same token a forest or ploughed field may be quite dark in tone because the surface scatters light away from the camera. Water bodies provide a special case in that they often appear to be black because they are 'transparent' and absorb rather than reflect light. An exception occurs when the incident angle of the sun's rays is low and the camera lies in the path of the reflected rays. In this case the water may act as a mirror, reflecting much of the incident light and appearing a very light and bright shade on the photograph. Calibration and 'ground truthing' are important to successful aerial photograph interpretation.

The appearance of various kinds of land surface on a photograph often is difficult to predict but types can be established in known areas and these then form the basis for interpretation elsewhere. For example, a known marsh can be used to establish the 'marsh type' surface and others can therefore be identified in unmapped areas nearby. At other times it may be necessary to conduct a field study to delineate a particular landform or landsurface type in order to correlate the real thing with the aerial photograph analogue.

Photogrammetry

What is Photogrammetry?

Photogrammetry is the science of making measurements from photographs.

The output of photogrammetry is typically a map, drawing, measurement, or a 3D model of some real-world object or scene. Many of the maps we use today are created with photogrammetry and photographs taken from aircraft.

Photogrammetry is the science behind the creation of almost every topographic map made since the 1930s. The human ability to perceive depth is the basis for the science and technology of photogrammetry. This ability to see in three dimensions is due to the offset in perspective centers between the left and the right eyes. Photographs taken to mimic this perspective shift are referred to as stereoscopic or stereo (meaning two). A stereo pair (or series) of images is taken from consecutive positions and overlap each other by at least 60%. Through the use of photogrammetry, highly detailed three-dimensional data can be derived from the two-dimensional photographs of a stereo pair.

The formal definition of photogrammetry is the art, science, and technology of obtaining reliable information about physical objects and the environment through the process of recording, measuring, and interpreting photographic images and patterns of electromagnetic radiant energy and other phenomena. In many instances, the use of photogrammetry can be more efficient, less labour-intensive, and more cost effective than other types of field, 3D data collection, resulting in products that have a level of detail, accuracy, range, and price that is difficult to match with other technologies (Birch 2006). As described previously, the main component necessary for a photogrammetric project is a series of overlapping stereoscopic images. The stereo images may be captured by a large variety of cameras at almost any height or platform (from tripod to earth-orbiting satellite).

Photogrammetry is informally divided into two basic categories according to the height of the platform—traditional or aerial and nontraditional or close-range. Advances in commercially available and cost-effective photogrammetric software, high-resolution digital cameras, high-performance laptop computers, and unmanned aerial systems (UAS) have caused those categories to overlap in recent years. This document refers to the traditional process of acquiring and using large-format, vertical (film or digital) aerial images as aerial photogrammetry. Close range is used to refer to photographs with an object-to-camera distance of less than 300 m.

Types of Photogrammetry

Photogrammetry can be classified a number of ways but one standard method is to split the field based on camera location during photography. On this basis we have

- Aerial Photogrammetry,
- Close-Range Photogrammetry

In Aerial Photogrammetry the camera is mounted in an aircraft and is usually pointed vertically towards the ground. Multiple overlapping photos of the ground are taken as the aircraft flies along a flight path. These photos are processed in a stereoplotter (an instrument that lets an operator see two photos at once in a stereo view). These photos are also used in automated processing for Digital Elevation Model (DEM) creation.

Aerial Photogrammetry

Aerial Photogrammetry utilizes large-format imagery and ground coordinate information to effectively recreate the geometry of a portion of the earth in a virtual environment. In this virtual environment, reliable horizontal and vertical measurements can be made and recorded (or compiled) directly into a geospatial data file. Accurate measurements can be recorded from aerial photographic images, by using traditional methods, only when the following conditions are met: a) stereoscopic image pairs (two or more overlapping photographs) cover the object to be analyzed; b) accurate *x*, *y*, and *z* coordinates are known for at least three defined object points in the overlapping photographs; and c) a calibrated mapping or metric camera is used to take the photographs.

The compilation of planimetric features (such as roads and streams) and topographic information (such as digital terrain models [DTM] and topographic contours) from the photographic sources is accomplished through the use of digital stereoscopic instruments. Digital, or softcopy, photogrammetric workstations require specialized software and hardware for viewing a pair of stereo images. In this virtual environment, an experienced operator can link the images with the ground control to collect precise horizontal and vertical coordinates for a point, line, polygon, or surface. The photogrammetric workstation recreates the geometry of the field subject through a series of mathematical operations. These procedures require a high level of expertise and repetition to maintain the operator's skill. The softcopy instrument has analytical capabilities to a submillimeter level. Thus, high accuracy ground control coordinate positions are needed to fully exploit the analytical capabilities of these instruments (Matthews et al. 2006).

Aerial Photogrammetry Workflow

Most large-format aerial photography is acquired through commercial contractors and is available in hard copy (film) or digital formats. With the exception of the BLM Alaska State Office, the Bureau does not have the capability to acquire large-format aerial imagery without the assistance of a commercial contractor. In most instances, it is best to work through photogrammetric specialists, such as the staff of the Division of Resource Services (DRS) at the NOC to ensure that the technical and user requirements are met. It is important to begin planning for the acquisition of imagery early in the project. Many elements need to be considered to ensure that project requirements are met (e.g., size of the area, terrain, vegetation cover, sun angle, type of sensor, and product accuracy).

Generally, aerial acquisition is designed and planned according to the specifications needed to generate a particular product with a specific scale and accuracy. The desired resolution dictates the height at which the contractor must fly to acquire the imagery. The following equation states the relation between scale and flying height of the aircraft (Hussain and Bethel 2004).

Scale = focal length

Flying height above terrain

For example; when film-based aerial photography is flown at a height of 305 m (1,000 feet) above mean terrain with a 153-mm (6-inch) focal length lens, the resulting photography has a scale of 1:2,000. With the resolving power of modern natural color film, the smallest object that can be detected is 5 cm. The distance of the sensor from the subject in commercial aerial photography is restricted by the altitude at which the plane can operate safely. In addition, the Federal Aviation Administration (FAA) imposes a minimum altitude limit of 1,000 feet above populated areas.

Many land management agencies, including the BLM, maintain collections of historical aerial photography of the lands they manage. The National Agriculture Imagery Program (NAIP) is managed by the U.S. Department of Agriculture Farm Services Agency (FSA) and acquires imagery of agricultural lands in the United States. Other governmental agencies work with the FSA to produce statewide coverage of natural-color, 1-, 2-, and 10-m resolution imagery that has been corrected for terrain and image distortions and is geospatially correct (orthorectified) on a cyclical basis. Information on these archives is available through individual agency websites. The BLM's aerial photography archive is maintained at the NOC.

Traditional, large-format (9-inch × 9-inch) aerial photography is film-based and is available in black and white, natural color, or color infrared formats depending on the project requirements. Each aerial camera has its own camera calibration report. This is a report generated with the results of a series of tests developed to reveal the geometric characteristics of the camera and lens. The photogrammetrist uses this information in subsequent calcu-lations to remove distortions, which helps to ensure a high-quality product. However, the imaging and mapping industry has started to migrate to a fully digital process in recent years. Often, traditional film is not used, as the imagery is captured in a digital format.

Presently, several sensor systems are able to capture aerial images digitally. The frame digital sensor is similar to a film camera in that it captures single frames. A push broom sensor (an array of several thousand charge coupled device [CCD] sensors) continuously collects data and then divides the data into frames. Digital images can also be captured in panchro-matic, multispectral, and infrared modes simultaneously. This is advantageous because the images may be captured concurrently and do not require additional flights. Digital sensors also offer capabilities that allow the photogrammetrist to enhance portions of the image through image processing techniques to compensate for adverse conditions such as shadows or high-contrast areas. This allows data collection in areas that might be obscured on film.

In Close range Photogrammetry the camera is close to the subject and is typically hand-held or on a tripod (but can be on a vehicle too). Usually this type of photogrammetry is non-topographic - that is, the output is not topographic products like terrain models or topographic maps, but instead drawings, 3D models, measurements and point clouds. Everyday cameras are used to model and measure buildings, engineering structures, forensic and accident scenes, mines, earth-works, stock-piles, archaeological artifacts, film sets, etc. This type of photogrammetry (CRP for short) is also sometimes called Image-Based Modeling.

The same basic principles of traditional aerial photogrammetry can be applied to stereoscopic pictures taken from lower altitudes or from the ground. Terrestrial, ground-based, and close-range are all descriptive terms that refer to photos taken with an object-to-camera distance of less than 300 m (1,000 feet). This distance equates to the minimum safe flying height above populated areas, as required by the FAA. Since the same basic principles apply to photographs taken from a camera mounted on a tripod (terrestrial) or suspended from a light sport aircraft (low-level aerial), both types of nontraditional photogrammetry are referred to in this report as close range photogrammetry (CRP).

A variety of cameras and platforms may be used to obtain the photographic images to be used in CRP processing, including cameras housed in unoccupied airborne vehicles, suspended below helium-filled blimps, and mounted on tripods. Through the use of these nontraditional methods, a resolution or ground sample distance of 0.25 mm and a spatial accuracy equivalent to 0.025 mm can be achieved. Theoretically, there is no limit to the resolution that can be achieved from CRP images (Matthews et al. 2006).

The BLM's national center in Denver has used close-range photogrammetric tech-niques to document resources since the late 1980s. At that time, although producing high-quality results, the close-range photogrammetric process could be tedious and time-consuming mainly because of the need to apply traditional techniques, workflow, and equipment to close-range image capture and processing. However, recent advances in commercially available and cost effective three dimensional measuring and modeling (3DMM) software, high-resolution digital cameras, and high performance laptop computers have revolutionized the CRP process. With these advances, the photogrammetric process can be moved from the laboratory environment into the field, and taken strictly from the hands of the photogrammetry expert and moved to the hands of the resource specialist.

Close Range Photogrammetry Workflow

As with traditional photogrammetry, the first step in the close range project workflow is establishing project requirements. These requirements include the area to be covered, type of ground cover, access to areas surrounding the subject, weather, accuracy of the final product, format of the final product, public visitation at the site, and others. However, with the relatively recent availability of reasonably priced, high quality digital cameras and 3DMM software, many of the traditional requirements in the photogrammetric process can be accomplished in nontraditional ways, providing a high level of flexibility.

An example of the increased flexibility can be seen in image acquisition. In most situations, a person who is experienced with CRP (such as the staff at the NOC), using a high-resolution commercial digital single-lens reflex (SLR) camera, can successfully capture ground-based photographs for a variety of projects. Most persons, following a few simple rules and taking good quality photographs, can capture the images necessary for completing a successful CRP project.

Project Design

Once project requirements have been determined, the process of project planning can begin. Accuracy, photographic scale, and project size are directly related. Typically, the higher the level of accuracy required, the larger the scale of photography necessary, which in turn increases the magnitude of the job. CRP projects can be placed into one of two categories. These categories are small (simple) and large (complex).

Small projects are those for which the area of interest is equivalent to about 5 m2 or smaller. This type of project may also be referred to as extreme close range, as a high precision of data points is expected. Although proper planning strengthens any project, the small terrestrial project may afford greater flexibility in the plan-ning process. Project design and image acquisition for projects of this scope may be successfully carried out by the resource specialists in the field. By following a set of standard operating procedures, successful completion of the capture of imagery for an extreme close-range project may be accomplished.

Large projects are those for which the area of interest is greater than 5 m^2 or where complexities of terrain or access add to the level of expertise needed for planning and execution of the project. Projects that fall into this category may require a preliminary site visit or the inspection of overview photographs taken of the area. A preliminary reconnaissance of the area will determine if close range photogrammetry is the best method for data capture, and if so, estimate resources necessary to accomplish the project. Also at this time, the data requirements and expectations of the end user can be discussed. Projects of this scope are best accomplished by those who have experience with close-range image capture and processing.

Sometimes the needed vantage points to capture imagery may not be present, necessitating alternatives to ground based capture. In these situations, a large ladder, aerial blimp, UAS, or other type of elevated platform may be needed. For these reasons, it is advantageous to contact the photogrammetric specialist for consultation early in project development to help with such items as budget, scheduling, site visit input, and others.

❑❑❑

3

Satellite Remote Sensing

Digital image processing is the use of computer algorithms to perform image processing on digital images. As a subcategory or field of digital signal processing, digital image processing has many advantages over analog image processing. It allows a much wider range of algorithms to be applied to the input data and can avoid problems such as the buildup of noise and signal distortion during processing. Since images are defined over two dimensions (perhaps more) digital image processing may be modeled in the form of multidimensional systems. Introduction to image processing

Image processing is a method to convert an image into digital form and perform some operations on it, in order to get an enhanced image or to extract some useful information from it. It is a type of signal dispensation in which input is image, like video frame or photograph and output may be image or characteristics associated with that image. Usually image processing system includes treating images as two dimensional signals while applying already set signal processing methods to them.

Image processing is a method to perform some operations on an image, in order to get an enhanced image or to extract some useful information from it. It is a type of signal processing in which input is an image and output may be image or characteristics/features associated with that image. Now-a-days, image processing is among rapidly growing technologies. It forms core research area within engineering and computer science disciplines too.

Image processing basically includes the following three steps.

- Importing the image with optical scanner or by digital photography.
- Analyzing and manipulating the image which includes data compression and image enhancement and spotting patterns that are not to human eyes like satellite photographs.

- Output is the last stage in which result can be altered image or report that is based on image analysis.

There are two types of methods used for image processing namely, analogue and digital image processing. Analogue image processing can be used for the hard copies like printouts and photographs. Image analysts use various fundamentals of interpretation while using these visual techniques. Digital image processing techniques help in manipulation of the digital images by using computers. The three general phases that all types of data have to undergo while using digital technique are pre-processing, enhancement, and display, information extraction.

In this lecture we will talk about a few fundamental definitions such as image, digital image, and digital image processing. Different sources of digital images will be discussed and examples for each source will be provided. The continuum from image processing to computer vision will be covered in this lecture. Finally we will talk about image acquisition and different types of image sensors.

Purpose of Image processing

The purpose of image processing is divided into 5 groups. They are:

1. Visualization - Observe the objects that are not visible.
2. Image sharpening and restoration - To create a better image.
3. Image retrieval - Seek for the image of interest.
4. Measurement of pattern – Measures various objects in an image.
5. Image recognition – Distinguish the objects in an image.

Types

The two types of **methods used for Image Processing** or **Analog and Digital** Image Processing. Analog or visual techniques of image processing can be used for the hard copies like printouts and photographs. Image analysts use various fundamentals of interpretation while using these visual techniques. The image processing is not just confined to area that has to be studied but on knowledge of analyst. Association is another important tool in image processing through visual techniques. So analysts apply a combination of personal knowledge and collateral data to image processing.

Digital Processing techniques help in manipulation of the digital images by using computers. As raw data from imaging sensors from satellite platform contains deficiencies. To get over such flaws and to get originality of information, it has to undergo various phases of processing. The three general phases that all types of data have to undergo while using digital technique are Pre-processing, enhancement and display, information extraction.

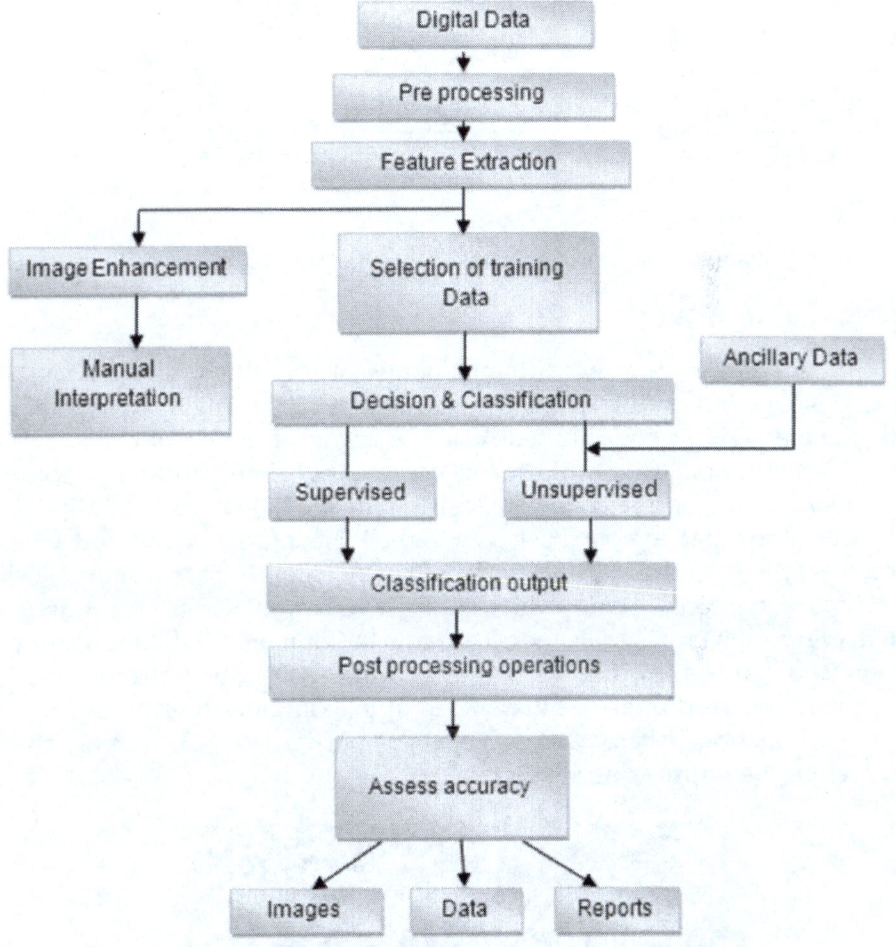

Figure 3.1 Digital data flow chart

Digital Image

A digital remotely sensed image is typically composed of picture elements (pixels) located at the intersection of each row i and column j in each K bands of imagery. Associated with each pixel is a number known as Digital Number (DN) or Brightness Value (BV) that depicts the average radiance of a relatively small area within a scene. A smaller number indicates low average radiance from the area and the high number is an indicator of high radiant properties of the area. The size of this area effects the reproduction of details within the scene. As pixel size is reduced more scene detail is presented in digital representation.

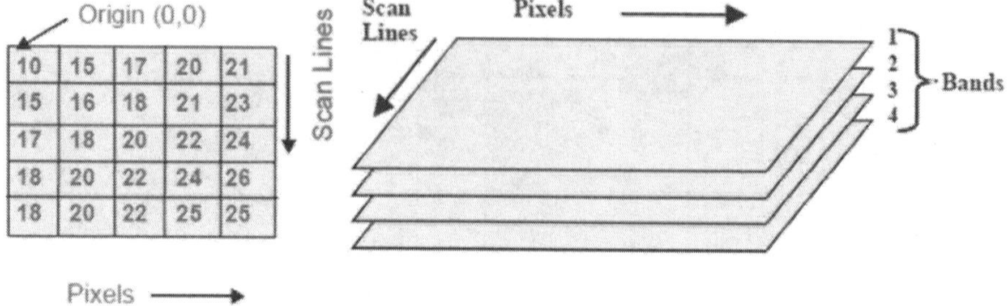

Figure 3.2 Sources of a digital image and multispectral image

Color Composites

While displaying the different bands of a multispectral data set, images obtained in different bands are displayed in image planes (other than their own) the color composite is regarded as False Color Composite (FCC). High spectral resolution is important when producing color components. For a true color composite an image data used in red, green and blue spectral region must be assigned bits of red, green and blue image processor frame buffer memory. A color infrared composite 'standard false color composite' is displayed by placing the infrared, red, green in the red, green and blue frame buffer memory. In this healthy vegetation shows up in shades of red because vegetation absorbs most of green and red energy but reflects approximately half of incident Infrared energy. Urban areas reflect equal portions of NIR, R & G, and therefore they appear as steel grey.

Figure 3.3 FCC Colour Image (See colour version on page 171)

Colour Composites : Spectral

'False Colour' composite (FCC)

IR band on Red

Red band on Green

Green band on Blue

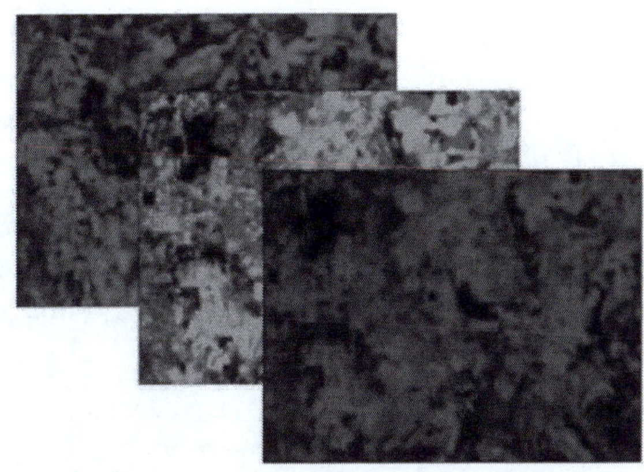

Figure 3.4 Colour composites (See colour version on page 171)

Image Rectification and Registration

Geometric distortions manifest themselves as errors in the position of a pixel relative to other pixels in the scene and with respect to their absolute position within some defined map projection. If left uncorrected, these geometric distortions render any data extracted from the image useless. This is particularly so if the information is to be compared to other data sets, be it from another image or a GIS data set. Distortions occur for many reasons.

For instance distortions occur due to changes in platform attitude (roll, pitch and yaw), altitude, earth rotation, earth curvature, panoramic distortion and detector delay. Most of these distortions can be modeled mathematically and are removed before you buy an image. Changes in attitude however can be difficult to account for mathematically and so a procedure called image rectification is performed. Satellite systems are however geometrically quite stable and geometric rectification is a simple procedure based on a mapping transformation relating real ground coordinates, say in easting and northing, to image line and pixel coordinates.

Rectification is a process of geometrically correcting an image so that it can be represented on a planar surface , conform to other images or conform to a map (Fig. 3.5). That is, it is the process by which geometry of an image is made planimetric. It is necessary when accurate area, distance and direction measurements are required to be made from the imagery. It is achieved by transforming the data from one grid system into another grid system using a geometric transformation.

Rectification is not necessary if there is no distortion in the image. For example, if an image file is produced by scanning or digitizing a paper map that is in the desired projection system, then that image is already planar and does not require rectification unless there is some skew or rotation of the image. Scanning and digitizing produce images that are planar, but do not contain any map coordinate information. These images need only to be geo-referenced, which is a much simpler process than rectification. In many cases, the image header can simply be updated with new map coordinate information. This involves redefining the map coordinate of the upper left corner of the image and the cell size (the area represented by each pixel).

Ground Control Points (GCP) are the specific pixels in the input image for which the output map coordinates are known. By using more points than necessary to solve the transformation equations a least squares solution may be found that minimises the sum of the squares of the errors. Care should be exercised when selecting ground control points as their number, quality and distribution affect the result of the rectification.

Once the mapping transformation has been determined a procedure called resampling is employed. Resampling matches the coordinates of image pixels to their real world coordinates and writes a new image on a pixel by pixel basis. Since the grid of pixels in the source image rarely matches the grid for the reference image, the pixels are resampled so that new data file values for the output file can be calculated.

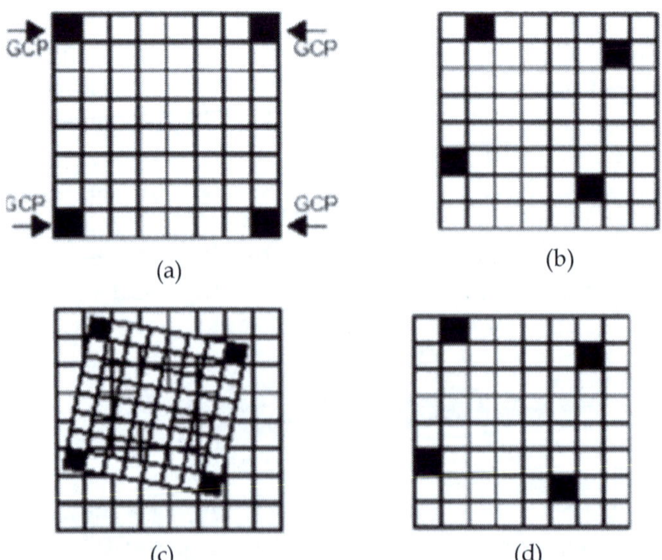

Figure 3.5 Image Rectification (a & b) Input and reference image with GCP locations, (c) using polynomial equations the grids are fitted together, (d) using resampling method the output grid pixel values are assigned (*Source* modified from ERDAS Field guide)

Image Enhancement Techniques

Image enhancement techniques improve the quality of an image as perceived by a human. These techniques are most useful because many satellite images when examined on a colour display give inadequate information for image interpretation. There is no conscious effort to improve the fidelity of the image with regard to some ideal form of the image. There exists a wide variety of techniques for improving image quality. The contrast stretch, density slicing, edge enhancement, and spatial filtering are the more commonly used techniques. Image enhancement is attempted after the image is corrected for geometric and radiometric distortions. Image enhancement methods are applied separately to each band of a multispectral image. Digital techniques have been found to be most satisfactory than the photographic technique for image enhancement, because of the precision and wide variety of digital processes.

Contrast

Contrast generally refers to the difference in luminance or grey level values in an image and is an important characteristic. It can be defined as the ratio of the maximum intensity to the minimum intensity over an image. Contrast ratio has a strong bearing on the resolving power and detectability of an image. Larger this ratio, more easy it is to interpret the image. Satellite images lack adequate contrast and require contrast improvement.

Contrast Enhancement

Contrast enhancement techniques expand the range of brightness values in an image so that the image can be efficiently displayed in a manner desired by the analyst. The density values in a scene are literally pulled farther apart, that is, expanded over a greater range. The effect is to increase the visual contrast between two areas of different uniform densities. This enables the analyst to discriminate easily between areas initially having a small difference in density.

Linear Contrast Stretch

This is the simplest contrast stretch algorithm. The grey values in the original image and the modified image follow a linear relation in this algorithm. A density number in the low range of the original histogram is assigned to extremely black and a value at the high end is assigned to extremely white. The remaining pixel values are distributed linearly between these extremes. The features or details that were obscure on the original image will be clear in the contrast stretched image. Linear contrast stretch operation can be represented graphically as shown in given below.

To provide optimal contrast and colour variation in colour composites the small range of grey values in each band is stretched to the full brightness range of the output or display unit.

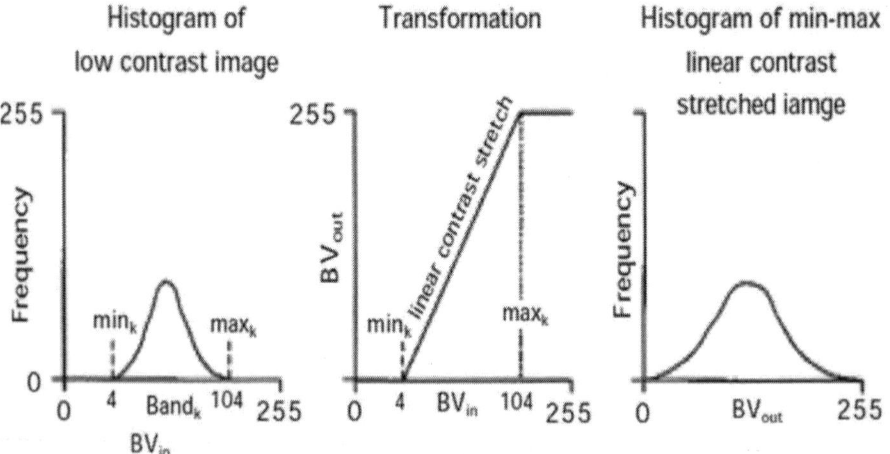

Figure 3.6 Linear Contrast Stretch algorithm

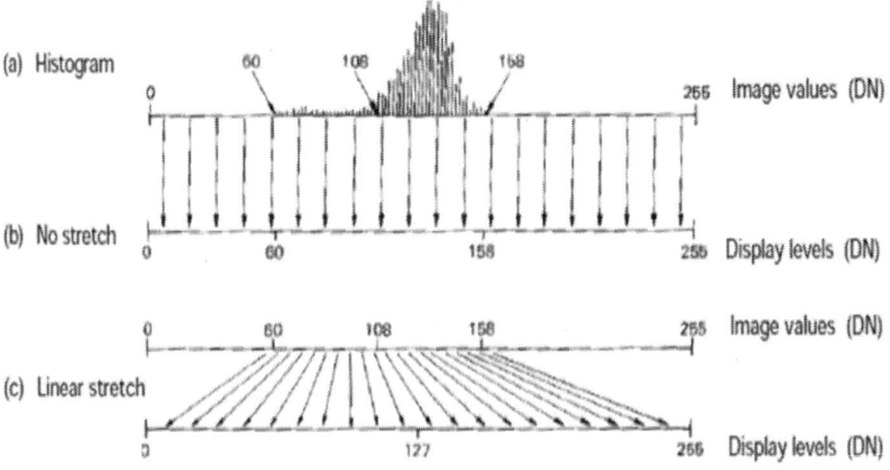

Figure 3.7 Linear Contrast Stretch (*Source:* Lillesand and Kiefer, 1993)

Non-Linear Contrast Enhancement

In these methods, the input and output data values follow a non-linear transformation. The general form of the nonlinear contrast enhancement is defined by y = f (x), where x is the input data value and y is the output data value. The non-linear contrast enhancement techniques have been found to be useful for enhancing the colour contrast between the nearly classes and subclasses of a main class.

A type of non linear contrast stretch involves scaling the input data logarithmically. This enhancement has greatest impact on the brightness values found in the darker part of histogram. It could be reversed to enhance values in brighter part of histogram by scaling the input data using an inverse log function. Histogram equalization is another non-linear contrast enhancement technique. In this technique, histogram of the original image is redistributed to produce a uniform population density. This is obtained by grouping certain adjacent grey values. Thus the number of grey levels in the enhanced image is less than the number of grey levels in the original image.

Spatial Filtering

A characteristic of remotely sensed images is a parameter called spatial frequency defined as number of changes in Brightness Value per unit distance for any particular part of an image. If there are very few changes in Brightness Value once a given area in an image, this is referred to as low frequency area. Conversely, if the Brightness Value changes dramatically over short distances, this is an area of high frequency.

Spatial filtering is the process of dividing the image into its constituent spatial frequencies, and selectively altering certain spatial frequencies to emphasize some image features. This technique increases the analyst's ability to discriminate detail. The three types of spatial filters used in remote sensor data processing are : Low pass filters, Band pass filters and High pass filters.

Low-Frequency Filtering in the Spatial Domain

Image enhancements that deemphasize or block the high spatial frequency detail are low-frequency or low-pass filters. The simplest low-frequency filter evaluates a particular input pixel brightness value, BVin, and the pixels surrounding the input pixel, and outputs a new brightness value, BVout, that is the mean of this convolution. The size of the neighbourhood convolution mask or kernel (n) is usually 3x3, 5x5, 7x7, or 9x9. The simple smoothing operation will, however, blur the image, especially at the edges of objects. Blurring becomes more severe as the size of the kernel increases.

Using a 3x3 kernel can result in the low-pass image being two lines and two columns smaller than the original image. Techniques that can be applied to deal with this problem include (1) artificially extending the original image beyond its border by repeating the original border pixel brightness values or (2) replicating the averaged brightness values near the borders, based on the image behaviour within a view pixels of the border. The most commonly used low pass filters are mean, median and mode filters.

High-Frequency Filtering in the Spatial Domain

High-pass filtering is applied to imagery to remove the slowly varying components and enhance the high-frequency local variations. Brightness values tend to be highly correlated in a nine-element window. Thus, the high frequency filtered image will have a relatively narrow intensity histogram. This suggests that the output from most high frequency filtered images must be contrast stretched prior to visual analysis.

Edge Enhancement in the Spatial Domain

For many remote sensing earth science applications, the most valuable information that may be derived from an image is contained in the edges surrounding various objects of interest. Edge enhancement delineates these edges and makes the shapes and details comprising the image more conspicuous and perhaps easier to analyze. Generally, what the eyes see as pictorial edges are simply sharp changes in brightness value between two adjacent pixels. The edges may be enhanced using either linear or nonlinear edge enhancement techniques.

Linear Edge Enhancement

A straightforward method of extracting edges in remotely sensed imagery is the application of a directional first difference algorithm and approximates the first derivative between two adjacent pixels. The algorithm produces the first difference of the image input in the horizontal, vertical, and diagonal directions.

The Laplacian operator generally highlights point, lines, and edges in the image and suppresses uniform and smoothly varying regions. Human vision physiological research suggests that we see objects in much the same way. Hence, the use of this operation has a more natural look than many of the other edge-enhanced images.

Band Rationing

Sometimes differences in brightness values from identical surface materials are caused by topographic slope and aspect, shadows, or seasonal changes in sunlight illumination angle and intensity. These conditions may hamper the ability of an interpreter or classification algorithm to identify correctly surface materials or land use in a remotely sensed image. Fortunately, ratio transformations of the remotely sensed data can, in certain instances, be applied to reduce the effects of such environmental conditions. In addition to minimizing the effects of environmental factors, ratios may also provide unique information not available in any single band that is useful for discriminating between soils and vegetation.

The mathematical expression of the ratio function is

$$BVi,j,r = BVi,j,k/BVi,j.l$$

Where BVi,j,r is the output ratio value for the pixel at rwo, i, column j; BVi,j,k is the brightness value at the same location in band k, and BVi,j,l is the brightness value in band l. Unfortunately, the computation is not always simple since BVi,j = 0 is possible. However, there are alternatives. For example, the mathematical domain of the function is 1/255 to 255 (i.e., the range of the ratio function includes all values beginning at 1/255, passing through 0 and ending at 255). The way to overcome this problem is simply to give any BVi,j with a value of 0 the value of 1.

Ratio images can be meaningfully interpreted because they can be directly related to the spectral properties of materials. Ratioing can be thought of as a method of enhancing minor differences between materials by defining the slope of spectral curve between two bands. We must understand that dissimilar materials having similar spectral slopes but different albedos, which are easily separable on a standard image, may become inseparable on ratio images. Given below Figure shows a situation where Deciduous and Coniferous Vegetation crops out on both the sunlit and shadowed sides of a ridge.

In the individual bands the reflectance values are lower in the shadowed area and it would be difficult to match this outcrop with the sunlit outcrop. The ratio values, however, are nearly identical in the shadowed and sunlit areas and the sandstone outcrops would have similar signatures on ratio images. This removal of illumination differences also eliminates the dependence of topography on ratio images.

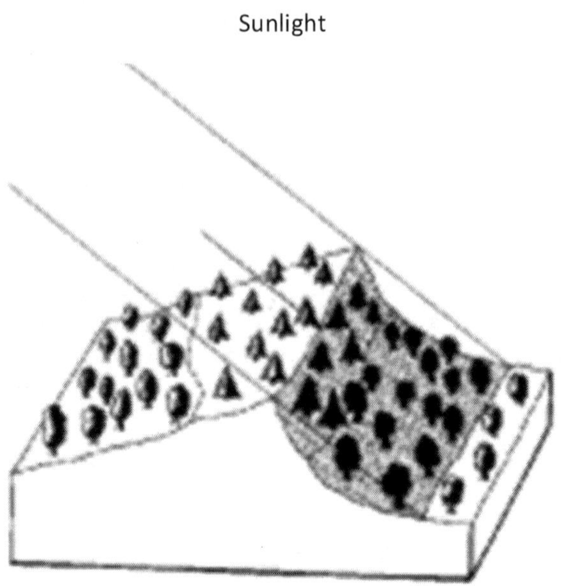

Figure 3.8 Reduction of Scene Illumination effect through spectral ratioing (*Source:* Lillesand & Kiefer, 1993)

Table 3.1 Land cover band Ratio

Landcover/ Illumination	Digital Number		Ratio
	Band A	Band B	
Deciduous			
Sunlit	48	50	.96
Shadow	18	19	.95
Coniferous			
Sunlit	31	45	.69
Shadow	11	16	.69

Principal Component Analysis

The multispectral image data is usually strongly correlated from one band to the other. The level of a given picture element on one band can to some extent be predicted from the level of that same pixel in another band. Principal component analysis is a pre-processing transformation that creates new images from the uncorrelated values of different images. This is accomplished by a linear transformation of variables that corresponds to a rotation and translation of the original coordinate system. Principal component analysis operates on all bands together. Thus, it alleviates the difficulty of selecting appropriate bands associated with the band rationing operation. Principal components describe the data more efficiently than the original band reflectance values. The first principal component accounts for a maximum portion of the variance in the data set, often as high as 98%. Subsequent principal components account for successively smaller portions of the remaining variance.

Principal component transformations are used for spectral pattern recognition as well as image enhancement. When used before pattern recognition, the least important principal components are dropped altogether. This permits us to omit the insignificant portion of our data set and thusavoids the additional computer time. The transformation functions are determined during the training stage. Principal component images may be analysed as separate black and white images, or any three component images may be colour coded to form a colour composite. Principal component enhancement techniques are particularly appropriate in areas where little a priori information concerning the region is available.

Image Fusion Techniques

The satellites cover different portions of the electromagnetic spectrum and record the incoming radiations at different spatial, temporal, and spectral resolutions. Most of these sensors operate in two modes: *multispectral* mode and the *panchromatic* mode. The *panchromatic* mode corresponds to the observation over a broad spectral band (similar to a typical black and white photograph) and the *multispectral* (color) mode corresponds to the observation in a number of relatively narrower bands. For example in the IRS - 1D, LISS III operates in the multispectral mode. It records energy in the green (0.52 h"$%þ 0.59 μm), red (0.62-0.68 μm), near infrared (0.77- 0.86 μm) and mid-infrared (1.55 h"$%þ 1.70 μm). In the same satellite PAN operates in the panchromatic mode. SPOT is another satellite, which has a combination of sensor operating in the multispectral and panchromatic mode. Above information is also expressed by saying that the multispectral mode has a better ***spectral resolution*** than the panchromatic mode.

Now coming to the ***spatial resolution***, most of the satellites are such that the *panchromatic* mode has a better *spatial resolution* than the *multispectral* mode, for e.g. in IRS -1C, PAN has a spatial resolution of 5.8 m whereas in the case of LISS it is 23.5 m. Better is the spatial resolution, more detailed information about a land use is present in the imagery, hence usually PAN data is used for observing and separating various feature. Both these type of sensors have their particular utility as per the need of user. If the need of the user is to separate two different kind of land uses, LISS III is used, whereas for a detailed map preparation of any area, PAN imagery is extremely useful. Image Fusion is the combination of two or more different images to form a new image (by using a certain algorithm).

The commonly applied Image Fusion Techniques are:

1. IHS Transformation
2. PCA
3. Brovey Transform
4. Band Substitution.

Image Classification

The overall objective of image classification is to automatically categorize all pixels in an image into land cover classes or themes. Normally, multispectral data are used to perform the classification, and the spectral pattern present within the data for each pixel is used as numerical basis for categorization. That is, different feature types manifest different combination of DNs based on their inherent spectral reflectance and emittance properties. The term *classifier* refers loosely to a computer program that implements a specific procedure for image classification. Over the years scientists have devised many classification strategies. From these alternatives the analyst must select the classifier that will best accomplish a specific task. At present it is not possible to state that a given classifier is "best" for all situations because characteristics of each image and the circumstances for each study vary so greatly. Therefore, it is essential that the analyst understands the alternative strategies for image classification.

The traditional methods of classification mainly follow two approaches: unsupervised and supervised. The unsupervised approach attempts spectral grouping that may have an unclear meaning from the user's point of view. Having established these, the analyst then tries to associate an information class with each group. The unsupervised approach is often referred to as clustering and results in statistics that are for spectral, statistical clusters. In the supervised approach to classification, the image analyst supervises the pixel categorization process by specifying to the computer algorithm; numerical descriptors of the various land cover types present in the scene. To do this, representative sample sites of known cover types, called training areas or training sites, are used to compile a numerical interpretation key that describes the spectral attributes for each feature type of interest. Each pixel in the data set is then compared numerically to each category in the interpretation key and labeled with the name of the category it looks most like. In the supervised approach the user defines useful information categories and then examines their spectral separability whereas in the unsupervised approach he first determines spectrally separable classes and then defines their informational utility.

It has been found that in areas of complex terrain, the unsupervised approach is preferable to the supervised one. In such conditions if the supervised approach is used, the user will have difficulty in selecting training sites because of the variability of spectral response within each class. Consequently, a prior ground data collection can be very time consuming. Also, the supervised approach is subjective in the sense that the analyst tries to classify information categories, which are often composed of several spectral classes whereas spectrally distinguishable classes will be revealed by the unsupervised approach, and hence ground data collection requirements may be reduced. Additionally, the unsupervised approach has the potential advantage of revealing discriminable classes unknown from previous work. However, when definition of representative training areas is possible and statistical information classes show a close correspondence, the results of supervised classification will be superior to unsupervised classification.

Unsupervised classification

Unsupervised classifiers do *not utilize* training data as the basis for classification. Rather, this family of classifiers involves algorithms that examine the unknown pixels in an image and aggregate them into a number of classes based on the natural groupings or clusters present in the image values. It performs very well in cases where the values within a given cover type are close together in the measurement space, data in different classes are comparatively well separated. The classes that result from unsupervised classification are spectral classes because they are based solely on the natural groupings in the image values, the identity of the spectral classes will not be initially known. The analyst must compare the classified data with some form of reference data (such as larger scale imagery or maps) to determine the identity and informational value of the spectral classes. In the supervised approach we define useful information categories and then examine their spectral separability; in the unsupervised approach we determine spectrally separable classes and then define their informational utility.

There are numerous clustering algorithms that can be used to determine the natural spectral groupings present in data set. One common form of clustering, called the "K-means" approach also called as ISODATA (Interaction Self Organizing Data Analysis Technique) accepts from the analyst the number of clusters to be located in the data. The algorithm then arbitrarily "seeds", or locates, that number of cluster centers in the multidimensional measurement space. Each pixel in the image is then assigned to the cluster whose arbitrary mean vector is closest. After all pixels have been classified in this manner, revised mean vectors for each of the clusters are computed. The revised means are then used as the basis of reclassification of the image data.

The procedure continues until there is no significant change in the location of class mean vectors between successive iterations of the algorithm. Once this point is reached, the analyst determines the land cover identity of each spectral class. Because the K-means approach is iterative, it is computationally intensive. Therefore, it is often applied only to image sub-areas rather than to full scenes.

Supervised classification

Supervised classification can be defined normally as the process of samples of known identity to classify pixels of unknown identity. Samples of known identity are those pixels located within training areas. Pixels located within these areas term the training samples used to guide the classification algorithm to assigning specific spectral values to appropriate informational class.

The basic steps involve in a typical supervised classification procedure are illustrated on Given Figure.

- The training stage
- Feature selection
- Selection of appropriate classification algorithm
- Post classification smoothening
- Accuracy assessment.

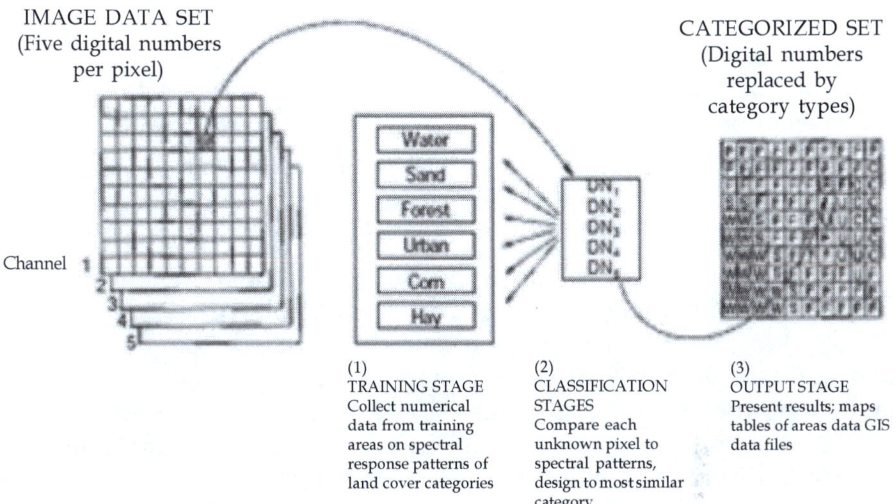

Figure 3.9 Basic Steps in Supervised Classification

Conclusions

Digital image processings of satellite data can be primarily grouped into three categories : Image Rectification and Restoration, Enhancement and Information extraction. Image rectification is the preprocessing of satellite data for geometric and radiometric connections. Enhancement is applied to image data in order to effectively display data for subsequent visual interpretation. Information extraction is based on digital classification and is used for generating digital thematic map.

Ground Truth Verification

Ground truth is a term used in various fields to refer to information provided by direct observation as opposed to information provided by inference.

In remote sensing, "ground truth" refers to information collected on location. Ground truth allows image data to be related to real features and materials on the ground. The collection of ground truth data enables calibration of remote sensing data, and aids in the interpretation and analysis of what is being sensed.

When a remote sensing or image processing technique is under development, it is mandatory to have sufficient ground truth data to test not only the accuracy of the final image analysis output, but also the intermediate steps in that process.

In remote sensing, "ground truth" refers to information collected on location. Ground truth allows image data to be related to real features and materials on the ground. The collection of ground-truth data enables calibration of remote-sensing data, and aids in the interpretation and analysis of what is being sensed. Examples include cartography, meteorology, analysis of aerial photographs, satellite imagery and other techniques in which data are gathered at a distance.

More specifically, ground truth may refer to a process in which a pixel on a satellite image is compared to what is there in reality (at the present time) in order to verify the contents of the pixel on the image. In the case of a classified image, it allows supervised classification to help determine the accuracy of the classification performed by the remote sensing software and therefore minimize errors in the classification such as errors of commission and errors of omission.

Ground truth is usually done on site, performing surface observations and measurements of various properties of the features of the ground resolution cells that are being studied on the remotely sensed digital image. It also involves taking geographic coordinates of the ground resolution cell with GPS technology and comparing those with the coordinates of the pixel being studied provided by the remote sensing software to understand and analyze the location errors and how it may affect a particular study.

Ground truth is important in the initial supervised classification of an image. When the identity and location of land cover types are known through a combination of field work, maps, and personal experience these areas are known as training sites. The spectral characteristics of these areas are used to train the remote sensing software using decision rules for classifying the rest of the image. These decision rules such as Maximum Likelihood Classification, Parallelepiped Classification, and Minimum Distance Classification offer different techniques to classify an image. Additional ground truth sites allow the remote sensor to establish an error matrix which validates the accuracy of the classification method used. Different classification methods may have different percentages of error for a given classification project. It is important that the remote sensor chooses a classification method that works best with the number of classifications used while providing the least amount of error.

Ground truth also helps with atmospheric correction. Since images from satellites obviously have to pass through the atmosphere, they can get distorted because of absorption in the atmosphere. So ground truth can help fully identify objects in satellite photos.

Errors of commission

An example of an error of commission is when a pixel reports the presence of a feature (such as trees) that, in reality, is absent (no trees are actually present). Ground truthing ensures that the error matrices have a higher accuracy percentage than would be the case if no pixels were ground truthed. This value is the inverse of the user's accuracy, i.e. Commission Error = 1 - user's accuracy.

Errors of omission

An example of an error of omission is when pixels of a certain thing, for example maple trees, are not classified as maple trees. The process of ground truthing helps to ensure that the pixel is classified correctly and the error matrices are more accurate. This value is the inverse of the producer's accuracy, i.e. Omission Error = 1 - producer's accuracy.

Global Positioning System (GPS)

What is GPS?

The Global Positioning System (GPS) is a satellite-based navigation system made up of a network of 24 satellites placed into orbit by the U.S. Department of Defense. GPS was originally intended for military applications, but in the 1980s, the government made the system available for civilian use. GPS works in any weather conditions, anywhere in the world, 24 hours a day. There are no subscription fees or setup charges to use GPS.

How it works?

GPS satellites circle the earth twice a day in a very precise orbit and transmit signal information to earth. GPS receivers take this information and use trilateration to calculate the user's exact location. Essentially, the GPS receiver compares the time a signal was transmitted by a satellite with the time it was received. The time difference tells the GPS receiver how far away the satellite is. Now, with distance measurements from a few more satellites, the receiver can determine the user's position and display it on the unit's electronic map. A GPS receiver must be locked on to the signal of at least 3 satellites to calculate a 2-D position (latitude and longitude) and track movement. With four or more satellites in view, the receiver can determine the user's 3-D position (latitude, longitude and altitude). Once the user's position has been determined, the GPS unit can calculate other information, such as speed, bearing, track, trip distance, distance to destination, sunrise and sunset time and more.

How accurate is GPS?

Today's GPS receivers are extremely accurate, thanks to their parallel multi-channel design. Our 12 parallel channel receivers are quick to lock onto satellites when first turned on, and they maintain strong locks, even in dense foliage or urban settings with tall buildings. Certain atmospheric factors and other sources of error can affect the accuracy of GPS receivers. Garmin GPS receivers are accurate to within 15 meters, on average.

Newer Garmin GPS receivers with WAAS (Wide Area Augmentation System) capability can improve accuracy to less than 3 meters on average. No additional equipment or fees are required to take advantage of WAAS. Users can also get better accuracy with Differential GPS (DGPS), which corrects GPS signals to within an average of 3 to 5 meters. The U.S. Coast Guard operates the most common DGPS correction service. This system consists of a network of towers that receive GPS signals and transmit a corrected signal by beacon transmitters. In order to get the corrected signal, users must have a differential beacon receiver and beacon antenna in addition to their GPS.

The GPS satellite system

The 24 satellites that make up the GPS space segment are orbiting the earth about 12,000 miles above us. They are constantly moving, making two complete orbits in less than 24 hours. These satellites are travelling at speeds of roughly 7,000 miles an hour.

GPS satellites are powered by solar energy. They have backup batteries onboard to keep them running in the event of a solar eclipse, when there's no solar power. Small rocket boosters on each satellite keep them flying in the correct path.

Here are some other interesting facts about the GPS satellites (also called NAVSTAR, the official U.S. Department of Defense name for GPS):

- The first GPS satellite was launched in 1978.
- A full constellation of 24 satellites was achieved in 1994.
- Each satellite is built to last about 10 years. Replacements are constantly being built and launched into orbit.
- A GPS satellite weighs approximately 2,000 pounds and is about 17 feet across with the solar panels extended.
- Transmitter power is only 50 Watts or less.

What's the signal?

GPS satellites transmit two low power radio signals, designated L1 and L2. Civilian GPS uses the L1 frequency of 1575.42 MHz in the UHF band. The signals travel by line of sight, meaning they will pass through clouds, glass and plastic but will not go through most solid objects such as buildings and mountains.

A GPS signal contains 3 different bits of information - a pseudorandom code, ephemeris data and almanac data. The pseudorandom code is simply an I.D. code that identifies which satellite is transmitting information. You can view this number on your Garmin GPS unit's satellite page, as it identifies which satellites it's receiving.

Ephemeris data, which is constantly transmitted by each satellite, contains important information about the status of the satellite (healthy or unhealthy), current date and time. This part of the signal is essential for determining a position.

The almanac data tells the GPS receiver where each GPS satellite should be at any time throughout the day. Each satellite transmits almanac data showing the orbital information for that satellite and for every other satellite in the system.

Sources of GPS signal errors

Factors that can degrade the GPS signal and thus affect accuracy include the following:

- Ionosphere and troposphere delays - The satellite signal slows as it passes through the atmosphere. The GPS system uses a built-in model that calculates an average amount of delay to partially correct for this type of error.
- Signal multipath - This occurs when the GPS signal is reflected off objects such as tall buildings or large rock surfaces before it reaches the receiver. This increases the travel time of the signal, thereby causing errors.
- Receiver clock errors - A receiver's built-in clock is not as accurate as the atomic clocks onboard the GPS satellites. Therefore, it may have very slight timing errors.
- Orbital errors - Also known as ephemeris errors, these are inaccuracies of the satellite's reported location.
- Number of satellites visible - The more satellites a GPS receiver can "see," the better the accuracy. Buildings, terrain, electronic

interference, or sometimes even dense foliage can block signal reception, causing position errors or possibly no position reading at all. GPS units typically will not work indoors, underwater or underground.

- Satellite geometry/shading - This refers to the relative position of the satellites at any given time. Ideal satellite geometry exists when the satellites are located at wide angles relative to each other. Poor geometry results when the satellites are located in a line or in a tight grouping.
- Intentional degradation of the satellite signal - Selective Availability (SA) is an intentional degradation of the signal once imposed by the U.S. Department of Defense. SA was intended to prevent military adversaries from using the highly accurate GPS signals. The government turned off SA in May 2000, which significantly improved the accuracy of civilian GPS receivers.

Structure of GPS

The current GPS consists of three major segments. These are the space segment (SS), a control segment (CS), and a user segment (US). The U.S. Air Force develops, maintains, and operates the space and control segments. GPS satellites broadcast signals from space, and each GPS receiver uses these signals to calculate its three dimensional location (latitude, longitude, and altitude) and the current time.

The space segment is composed of 24 to 32 satellites in medium Earth orbit and also includes the payload adapters to the boosters required to launch them into orbit. The control segment is composed of a master control station (MCS), an alternate master control station, and a host of dedicated and shared ground antennas and monitor stations. The user segment is composed of hundreds of thousands of U.S. and allied military users of the secure GPS Precise Positioning Service, and hundreds of millions of civil, commercial, and scientific users of the Standard Positioning Service (see GPS navigation devices).

Space segment

The space segment (SS) is composed of the orbiting GPS satellites, or Space Vehicles (SV) in GPS parlance. The GPS design originally called for 24 SVs, eight each in three approximately circular orbits, but this was modified to six orbital planes with four satellites each. The six orbit planes have approximately 55° inclination (tilt relative to the Earth's equator) and are separated by 60° right ascension of the ascending node (angle along the equator from a reference point to the orbit's intersection). The orbital period is one-half a sidereal day, i.e., 11 hours and 58 minutes so that the satellites pass over the same locations or almost the same locations every day. The orbits are arranged so that at least six satellites are always within line of sight from almost everywhere on the Earth's surface. The result of this objective is that the four satellites are not evenly spaced (90 degrees) apart within each orbit. In general terms, the angular difference between satellites in each orbit is 30, 105, 120, and 105 degrees apart, which sum to 360 degrees.

Orbiting at an altitude of approximately 20,200 km (12,600 mi); orbital radius of approximately 26,600 km (16,500 mi), each SV makes two complete orbits each sidereal day, repeating the same ground track each day. This was very helpful during development because even with only four satellites, correct alignment means all four are visible from one spot for a few hours each day. For military operations, the ground track repeat can be used to ensure good coverage in combat zones.

As of February 2016, there are 32 satellites in the GPS constellation, 31 of which are in use. The additional satellites improve the precision of GPS receiver calculations by providing redundant measurements. With the increased number of satellites, the constellation was changed to a non-uniform arrangement. Such an arrangement was shown to improve reliability and availability of the system, relative to a uniform system, when multiple satellites fail. About nine satellites are visible from any point on the ground at any one time (see animation at right), ensuring considerable redundancy over the minimum four satellites needed for a position.

Control segment

The control segment is composed of:

1. A master control station (MCS),
2. An alternate master control station,
3. Four dedicated ground antennas, and
4. Six dedicated monitor stations.

The MCS can also access U.S. Air Force Satellite Control Network (AFSCN) ground antennas (for additional command and control capability) and NGA (National Geospatial-Intelligence Agency) monitor stations. The flight paths of the satellites are tracked by dedicated U.S. Air Force monitoring stations in Hawaii, Kwajalein Atoll, Ascension

Island, Diego Garcia, Colorado Springs, Colorado and Cape Canaveral, along with shared NGA monitor stations operated in England, Argentina, Ecuador, Bahrain, Australia and Washington DC. The tracking information is sent to the Air Force Space Command MCS at Schriever Air Force Base 25 km (16 mi) ESE of Colorado Springs, which is operated by the 2nd Space Operations Squadron (2 SOPS) of the U.S. Air Force. Then 2 SOPS contacts each GPS satellite regularly with a navigational update using dedicated or shared (AFSCN) ground antennas (GPS dedicated ground antennas are located at Kwajalein, Ascension Island, Diego Garcia, and Cape Canaveral). These updates synchronize the atomic clocks on board the satellites to within a few nanoseconds of each other, and adjust the ephemeris of each satellite's internal orbital model. The updates are created by a Kalman filter that uses inputs from the ground monitoring stations, space weather information, and various other inputs.

Satellite maneuvers are not precise by GPS standards—so to change a satellite's orbit, the satellite must be marked *unhealthy*, so receivers don't use it. After the satellite maneuver, engineers track the new orbit from the ground, upload the new ephemeris, and mark the satellite healthy again.

The Operation Control Segment (OCS) currently serves as the control segment of record. It provides the operational capability that supports GPS users and keeps the GPS system operational and performing within specification.

OCS successfully replaced the legacy 1970s era mainframe computer at Schriever Air Force Base in September 2007. After installation, the system helped enable upgrades and provide a foundation for a new security architecture that supported U.S. armed forces. OCS will continue to be the ground control system of record until the new segment, Next Generation GPS Operation Control System (OCX), is fully developed and functional.

The new capabilities provided by OCX will be the cornerstone for revolutionizing GPS's mission capabilities, and enabling Air Force Space Command to greatly enhance GPS operational services to U.S. combat forces, civil partners and myriad domestic and international users.

The GPS OCX program also will reduce cost, schedule and technical risk. It is designed to provide 50% sustainment cost savings through efficient software architecture and Performance-Based Logistics. In addition, GPS OCX is expected to cost millions less than the cost to upgrade OCS while providing four times the capability.

The GPS OCX program represents a critical part of GPS modernization and provides significant information assurance improvements over the current GPS OCS program.

- OCX will have the ability to control and manage GPS legacy satellites as well as the next generation of GPS III satellites, while enabling the full array of military signals.
- Built on a flexible architecture that can rapidly adapt to the changing needs of today's and future GPS users allowing immediate access to GPS data and constellation status through secure, accurate and reliable information.
- Provides the warfighter with more secure, actionable and predictive information to enhance situational awareness.
- Enables new modernized signals (L1C, L2C, and L5) and has M-code capability, which the legacy system is unable to do.
- Provides significant information assurance improvements over the current program including detecting and preventing cyber attacks, while isolating, containing and operating during such attacks.
- Supports higher volume near real-time command and control capabilities and abilities.

On September 14, 2011, the U.S. Air Force announced the completion of GPS OCX Preliminary Design Review and confirmed that the OCX program is ready for the next phase of development.

The GPS OCX program has missed major milestones and is pushing the GPS IIIA launch beyond April 2016.

User segment

The user segment is composed of hundreds of thousands of U.S. and allied military users of the secure GPS Precise Positioning Service, and tens of millions of civil, commercial and scientific users of the Standard Positioning Service. In general, GPS receivers are composed of an antenna, tuned to the frequencies transmitted by the satellites, receiver-processors, and a highly stable clock (often a crystal oscillator). They may also include a display for providing location and speed information to the user. A receiver is often described by its number of channels: this signifies how many satellites it can monitor simultaneously. Originally limited to four or five, this has progressively increased over the years so that, as of 2007, receivers typically have between 12 and 20 channels.

GPS receivers may include an input for differential corrections, using the RTCM SC-104 format. This is typically in the form of an RS-232 port at 4,800 bit/s speed. Data is actually sent at a much lower rate, which limits the accuracy of the signal sent using RTCM. Receivers with internal DGPS receivers can outperform those using external RTCM data. As of 2006, even low cost units commonly include Wide Area Augmentation System (WAAS) receivers.

Many GPS receivers can relay position data to a PC or other device using the NMEA 0183 protocol. Although this protocol is officially defined by the National Marine Electronics Association (NMEA), references to this protocol have been compiled from public records, allowing open source tools like gpsd to read the protocol without violating intellectual property laws. Other proprietary protocols exist as well, such as the SiRF and MTK protocols. Receivers can interface with other devices using methods including a serial connection, USB, or Bluetooth.

❑❑❑

4

Global Positioning System

Introduction

The Global Positioning System (GPS) is a space-based navigation system that provides location and time information in all weather conditions, anywhere on or near the Earth where there is an unobstructed line of sight to four or more GPS satellites. The system provides critical capabilities to military, civil, and commercial users around the world. The United States government created the system, maintains it, and makes it freely accessible to anyone with a GPS receiver. The US began the GPS project in 1973 to overcome the limitations of previous navigation systems, integrating ideas from several predecessors, including a number of classified engineering design studies from the 1960s. The U.S. Department of Defense (DoD) developed the system, which originally used 24 satellites. It became fully operational in 1995. Roger L. Easton, Ivan A. Getting and Bradford Parkinson of the Applied Physics Laboratory are credited with inventing it.

Advances in technology and new demands on the existing system have now led to efforts to modernize the GPS and implement the next generation of GPS Block IIIA satellites and Next Generation Operational Control System (OCX). Announcements from Vice President Al Gore and the White House in 1998 initiated these changes. In 2000, the U.S. Congress authorized the modernization effort, GPS III. In addition to GPS, other systems are in use or under development. The Russian Global Navigation Satellite System (GLONASS) was developed contemporaneously with GPS, but suffered from incomplete coverage of the globe until the mid-2000s. There are also the planned European Union Galileo positioning system, China's BeiDou Navigation Satellite System, the Japanese Quasi-Zenith Satellite System, and India's Indian Regional Navigation Satellite System.

History of GPS

The design of GPS is based partly on similar ground-based radio-navigation systems, such as LORAN and the Decca Navigator, developed in the early 1940s and used by the British Royal Navy during World War II.

In 1956, the German-American physicist Friedwardt Winterberg proposed a test of general relativity – detecting time slowing in a strong gravitational field using accurate atomic clocks placed in orbit inside artificial satellites.

Special and general relativity predict that the clocks on the GPS satellites would be seen by the Earth's observers to run 38 microseconds faster per day than the clocks on the Earth. The GPS calculated positions would quickly drift into error, accumulating to 10 kilometers per day. The relativistic time effect of the GPS clocks running faster than the clocks on earth was corrected for in the design of GPS.

Predecessors

The Soviet Union launched the first man-made satellite, Sputnik 1, in 1957. Two American physicists, William Guier and George Weiffenbach, at Johns Hopkins's Applied Physics Laboratory (APL), decided to monitor Sputnik's radio transmissions. Within hours they realized that, because of the Doppler Effect, they could pinpoint where the satellite was along its orbit. The Director of the APL gave them access to their UNIVAC to do the heavy calculations required. The next spring, Frank McClure, the deputy director of the APL, asked Guier and Weiffenbach to investigate the inverse problem – pinpointing the user's location, given that of the satellite. (At the time, the Navy was developing the submarine-launched Polaris missile, which required them to know the submarine's location.) This led them and APL to develop the TRANSIT system. In 1959, ARPA (renamed DARPA in 1972) also played a role in TRANSIT.

The first satellite navigation system, TRANSIT, used by the United States Navy, was first successfully tested in 1960. It used a constellation of five satellites and could provide a navigational fix approximately once per hour. In 1967, the U.S. Navy developed the Timation satellite that proved the ability to place accurate clocks in space, a technology required by GPS.

In the 1970s, the ground-based OMEGA navigation system, based on phase comparison of signal transmission from pairs of stations, became the first worldwide radio navigation system. Limitations of these systems drove the need for a more universal navigation solution with greater accuracy.

While there were wide needs for accurate navigation in military and civilian sectors, almost none of those were seen as justification for the billions of dollars it would cost in research, development, deployment, and operation for a constellation of navigation satellites. During the Cold War arms race, the nuclear threat to the existence of the United States was the one need that did justify this cost in the view of the United States Congress. This deterrent effect is why GPS was funded. It is also the reason for the ultra secrecy at that time. The nuclear triad consisted of the United States Navy's submarine-launched ballistic missiles (SLBMs) along with United States Air Force (USAF) strategic bombers and intercontinental ballistic missiles (ICBMs). Considered vital to the nuclear deterrence posture, accurate determination of the SLBM launch position was a force multiplier.

Precise navigation would enable United States ballistic missile submarines to get an accurate fix of their positions before they launched their SLBMs. The USAF, with two thirds of the nuclear triad, also had requirements for a more accurate and reliable navigation system. The Navy and Air Force were developing their own technologies in parallel to solve what was essentially the same problem. To increase the survivability of ICBMs, there was a proposal to use mobile launch platforms (such as Russian SS-24 and SS-25) and so the need to fix the launch position had similarity to the SLBM situation.

In 1960, the Air Force proposed a radio-navigation system called MOSAIC (MObile System for Accurate ICBM Control) that was essentially a 3-D LORAN. A follow-on study, Project 57, was worked in 1963 and it was "in this study that the GPS concept was born." That same year, the concept was pursued as Project 621B, which had "many of the attributes that you now see in GPS"and promised increased accuracy for Air Force bombers as well as ICBMs.

Updates from the Navy TRANSIT system were too slow for the high speeds of Air Force operation. The Naval Research Laboratory continued advancements with their Timation (Time Navigation) satellites, first launched in 1967, and with the third one in 1974 carrying the first atomic clock into orbit. Another important predecessor to GPS came from a different branch of the United States military. In 1964, the United States Army orbited its first Sequential Collation of Range (SECOR) satellite used for geodetic surveying. The SECOR system included three ground-based transmitters from known locations that would send signals to the satellite transponder in orbit. A fourth ground-based station, at an undetermined position, could then use those signals to fix its location precisely. The last SECOR satellite was launched in 1969.

Decades later, during the early years of GPS, civilian surveying became one of the first fields to make use of the new technology, because surveyors could reap benefits of signals from the less-than-complete GPS constellation years before it was declared operational. GPS can be thought of as an evolution of the SECOR system where the ground-based transmitters have been migrated into orbit.

Development

With these parallel developments in the 1960s, it was realized that a superior system could be developed by synthesizing the best technologies from 621B, Transit, Timation, and SECOR in a multi-service program.

During Labor Day weekend in 1973, a meeting of about twelve military officers at the Pentagon discussed the creation of a *Defense Navigation Satellite System (DNSS)*. It was at this meeting that the real synthesis that became GPS was created. Later that year, the DNSS program was named *Navstar*, or Navigation System Using Timing and Ranging. With the individual satellites being associated with the name Navstar (as with the predecessors Transit and Timation), a more fully encompassing name was used to identify the constellation of Navstar satellites, *Navstar-GPS*. Ten "Block I" prototype satellites were launched between 1978 and 1985 (with one prototype being destroyed in a launch failure).

After Korean Air Lines Flight 007, a Boeing 747 carrying 269 people, was shot down in 1983 after straying into the USSR's prohibited airspace, in the vicinity of Sakhalin and Moneron Islands, President Ronald Reagan issued a directive making GPS freely available for civilian use, once it was sufficiently developed, as a common good. The first Block II satellite was launched on February 14, 1989, and the 24th satellite was launched in 1994. The GPS program cost at this point, not including the cost of the user equipment, but including the costs of the satellite launches, has been estimated at about US $5 billion (then-year dollars). Roger L. Easton is widely credited as the primary inventor of GPS.

Initially, the highest quality signal was reserved for military use, and the signal available for civilian use was intentionally degraded (Selective Availability). This changed with President Bill Clinton signing a policy directive in 1996 to turn off Selective Availability in May 2000 to provide the same precision to civilians that was afforded to the military. The directive was proposed by the U.S. Secretary of Defense, William

Perry, because of the widespread growth of differential GPS services to improve civilian accuracy and eliminate the U.S. military advantage. Moreover, the U.S. military was actively developing technologies to deny GPS service to potential adversaries on a regional basis.

Since its deployment, the U.S. has implemented several improvements to the GPS service including new signals for civil use and increased accuracy and integrity for all users, all the while maintaining compatibility with existing GPS equipment. Modernization of the satellite system has been an ongoing initiative by the U.S. Department of Defense through a series of satellite acquisitions to meet the growing needs of the military, civilians, and the commercial market.

As of early 2015, high-quality, FAA grade, Standard Positioning Service (SPS) GPS receivers provide horizontal accuracy of better than 3.5 meters, although many factors such as receiver quality and atmospheric issues can affect this accuracy.

GPS is owned and operated by the United States Government as a national resource. The Department of Defense is the steward of GPS. *Interagency GPS Executive Board (IGEB)* oversaw GPS policy matters from 1996 to 2004. After that the National Space-Based Positioning, Navigation and Timing Executive Committee was established by presidential directive in 2004 to advise and coordinate federal departments and agencies on matters concerning the GPS and related systems. The executive committee is chaired jointly by the deputy secretaries of defense and transportation. Its membership includes equivalent-level officials from the departments of state, commerce, and homeland security, the Joint Chiefs of Staff, and NASA. Components of the executive office of the president participate as observers to the executive committee, and the FCC chairman participates as a liaison.

The U.S. Department of Defense is required by law to "maintain a Standard Positioning Service (as defined in the federal radio navigation plan and the standard positioning service signal specification) that will be available on a continuous, worldwide basis," and "develop measures to prevent hostile use of GPS and its augmentations without unduly disrupting or degrading civilian uses."

What is GPS?

The Global Positioning System (GPS) is a satellite-based system that can be used to locate positions anywhere on the earth. Operated by the U.S. Department of Defense (DoD), NAVSTAR (NAVigation Satellite Timing and Ranging) GPS provides continuous (24 hours/day), real-time,

3-dimensional positioning, navigation and timing worldwide. Any person with a GPS receiver can access the system, and it can be used for any application that requires location coordinates.

The GPS system consists of three segments: 1) The space segment: the GPS satellites themselves, 2) The control system, operated by the U.S. military, and 3) The user segment, which includes both military and civilian users and their GPS equipment.

Space Segment: The GPS Constellation

The first GPS satellite was launched by the U.S. Air Force in early 1978. There are now at least 24 satellites orbiting the earth at an altitude of about 11,000 nautical miles. The high altitude insures that the satellite orbits are stable, precise and predictable, and that the satellites' motion through space is not affected by atmospheric drag. These 24 satellites make up a full GPS constellation.

The GPS satellites are powered primarily by sun-seeking solar panels, with nicad batteries providing secondary power. On board each GPS satellite are four atomic clocks, only one of which is in use at a time. These highly accurate atomic clocks enable GPS to provide the most accurate timing system that exists.

Satellite Orbit

There are four satellites in each of 6 orbital planes. Each plane is inclined 55 degrees relative to the equator, which means that satellites cross the equator tilted at a 55 degree angle. The system is designed to maintain full operational capability even if two of the 24 satellites fail.

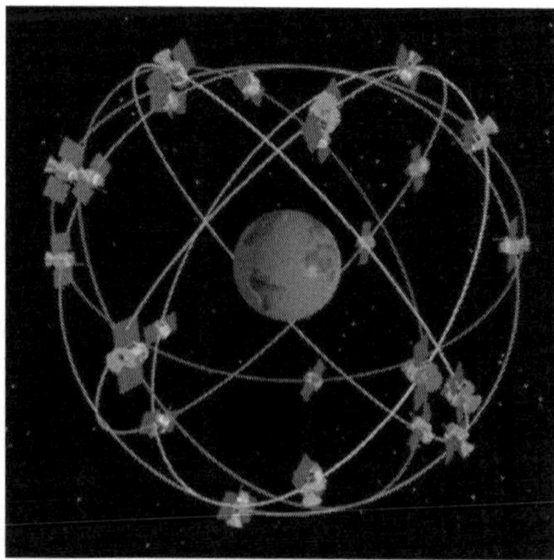

Figure 4.1 Satellite Orbits (See colour version on page 172)

GPS satellites complete an orbit in approximately 12 hours, which means that they pass over any point on the earth about twice a day. The satellites rise (and set) about four minutes earlier each day.

Satellite Signals

GPS satellites continuously broadcast satellite position and timing data via radio signals on two frequencies (L1 and L2). The radio signals travel at the speed of light (186,000 miles per second) and take approximately 6/100ths of a second to reach the earth.

The satellite signals require a direct line to GPS receivers and cannot penetrate water, soil, walls or other obstacles. For example, heavy forest canopy causes interference, making it difficult, if not impossible, to compute positions. In canyons (and "urban canyons" in cities) GPS signals are blocked by mountain ranges or buildings. If you place your hand over a GPS receiver antenna, it will stop computing positions.

Two kinds of code are broadcast on the L1 frequency (C/A code and P code). C/A (Coarse Acquisition) code is available to civilian GPS users and provides Standard Positioning Service (SPS). Using the Standard Positioning Service one can achieve 15 meter horizontal accuracy 95% of the time. This means that 95% of the time, the coordinates you read from your GPS receiver display will be within 15 meters of your true position on the earth. P (Precise) code is broadcast on both the L1 and L2 frequencies. P code, used for the Precise Positioning Service (PPS) is available only to the military. Using P code on both frequencies, a military receiver can achieve better accuracy than civilian receivers. Additional techniques can increase the accuracy of both C/A code and P code GPS receivers.

Control Segment: U.S. DoD Monitoring

The U.S. Department of Defense maintains a master control station at Falcon Air Force Base in Colorado Springs, CO. There are four other monitor stations located in Hawaii, Ascension Island, Diego Garcia and Kwajalein. The DoD stations measure the satellite orbits precisely. Any discrepancies between predicted orbits and actual orbits are transmitted back to the satellites. The satellites can then broadcast these corrections, along with the other position and timing data, so that a GPS receiver on the earth can precisely establish the location of each satellite it is tracking.

User Segment: Military and Civilian GPS Users

The U.S. military uses GPS for navigation, reconnaissance, and missile guidance systems. Civilian use of GPS developed at the same time as military uses were being established, and has expanded far beyond original expectations. There are civilian applications for GPS in almost every field, from surveying to transportation to natural resource management to agriculture. Most civilian uses of GPS, however, fall into one of four categories: navigation, surveying, mapping and timing.

GLONASS

The Russian government has developed a system, similar to GPS, called GLONASS. The first GLONASS satellite launch was in October 1982. The full constellation consists of 24 satellites in 3 orbit planes, which have a 64.8 degree inclination to the earth's equator. The GLONASS system now consists of 12 healthy satellites. GLONASS uses the same code for each satellite and many frequencies, whereas GPS which uses two frequencies and a different code for each satellite. Some GPS receiver manufacturers have incorporated the capability to receive both GPS and GLONASS signals. This increases the availability of satellites and the integrity of combined system.

GALILEO

Galileo is Europe's contribution to the next generation Global Navigation Satellite System (GNSS). Unlike GPS, which is funded by the public sector and operated by the U.S. Air Force, Galileo will be a civilcontrolled system that draws on both public and private sectors for funding. The service will be free at the point of use, but a range of chargeable services with additional features will also be offered. These additional features would include improved reception, accuracy and availability. Design of the Galileo system is being finalized and the delivery of initial services is targeted for 2008.

How can GPS be used?

GPS Applications in Agriculture

More and more producers today are using precision farming techniques that can help increase profits and protect the environment. Precision, or site-specific farming involves applying fertilizer, pesticides and other inputs only where they are needed. GPS-guided equipment is often used for variable rate application of fertilizer (based on soil tests) or pesticides (based on pest survey). GPS can also be used to develop the initial reference maps upon which variable rate applications are based. A GPS system on a combine with a yield monitor can be used to develop an on-the-go yield map or can be used to map weed locations from the combine when harvesting. Mounted in an airplane, GPS can be used to guide aerial spraying operations.

GPS can be used to locate weed, insect or diseases infestations and monitor their spread. It can also be used to navigate back to previously mapped infestations to apply controls. A field map can be created using GPS to record the coordinates of field borders, fence lines, canals, pipelines, and point locations such as wells, buildings, and landscape features. The resulting field map might be the first layer a producer would develop for an on-farm GIS (Geographic Information System). Additional layers showing crop damage from hail or drought, and riparian areas or wetlands could be mapped using GPS. Ranchers could use GPS to develop rangeland utilization maps and to navigate back to previously mapped areas or monitoring sites.

GPS Navigation: Land, Sea and Air

GPS is being used for emergency response (fire, ambulance, police), search and rescue, fleet management (trucking, delivery vehicles, and public transportation) and for automobile guidance systems. Recreational uses of GPS include navigation while hiking, hunting, or skiing. GPS is even used on golf courses to track golf carts, and to let players know how far it is to the center of the greens.

On our nation's waterways, GPS is being used for recreational sailing and fishing and for commercial shipping fleet management. Assisted steering, risk assessment and hazard warning systems for marine navigation are being developed using GPS.

In the air, GPS is being used for en-route navigation (helicopter, airplane, hot-air balloon), aircraft landing, and air-collision avoidance systems.

GPS Applications: Mapping and Surveying

GPS applications in natural resource management include inventory and mapping of soils, vegetation types, threatened and endangered species, lake and stream boundaries and wildlife habitat. GPS has been used to aid in damage assessment after natural disasters such as fires, floods and earthquakes. GPS has also been used to map archaeological sites and for infrastructure (streets, highways and utilities) mapping, management, and planning for future growth. Engineers use GPS for surveying when building roads, bridges and other structures.

Other GPS Applications: The Possibilities are Endless

Other uses of GPS include real estate valuation and taxation assessment, air quality studies, environmental protection, demographic analysis including marketing studies, atmospheric studies, oil and gas exploration, and scientific exploration. There are many additional current and possible uses for GPS. Any application where location information is needed is a possible candidate for GPS.

How does GPS work?

Calculating a Position

A GPS receiver calculates its position by a technique called satellite ranging, which involves measuring the distance between the GPS receiver and the GPS satellites it is tracking. The range (the range a receiver calculates is actually a pseudorange, or an estimate of range rather than a true range) or distance, is measured as elapsed transit time. The position of each satellite is *known*, and the satellites transmit their positions as part of the "messages" they send via radio waves. The GPS receiver on the ground is the *unknown* point, and must compute its position based on the information it receives from the satellites.

Measuring Distance to Satellites

The first step in measuring the distance between the GPS receiver and a satellite requires measuring the *time* it takes for the signal to travel from the satellite to the receiver. Once the receiver knows how much time has elapsed, it multiplies the travel time of the signal times the *speed of light* (because the satellite signals travel at the speed of light, approximately 186,000 miles per second) to compute the distance. Distance measurements to four satellites are required to compute a 3-dimensional (latitude, longitude and altitude) position.

In order to measure the travel time of the satellite signal, the receiver has to know when the signal left the satellite and when the signal reached the receiver. Knowing when the signal reaches the receiver is easy; the GPS receiver just "checks" its internal clock when the signal arrives to see what time it is. But how does it "know" when the signal left the satellite? All GPS receivers are synchronized with the satellites so they generate the same digital code at the same time. When the GPS receiver receives a code from a satellite, it can look back in its memory bank and "remember" when it emitted the same code. This little "trick" allows the GPS receiver to determine when the signal left the satellite.

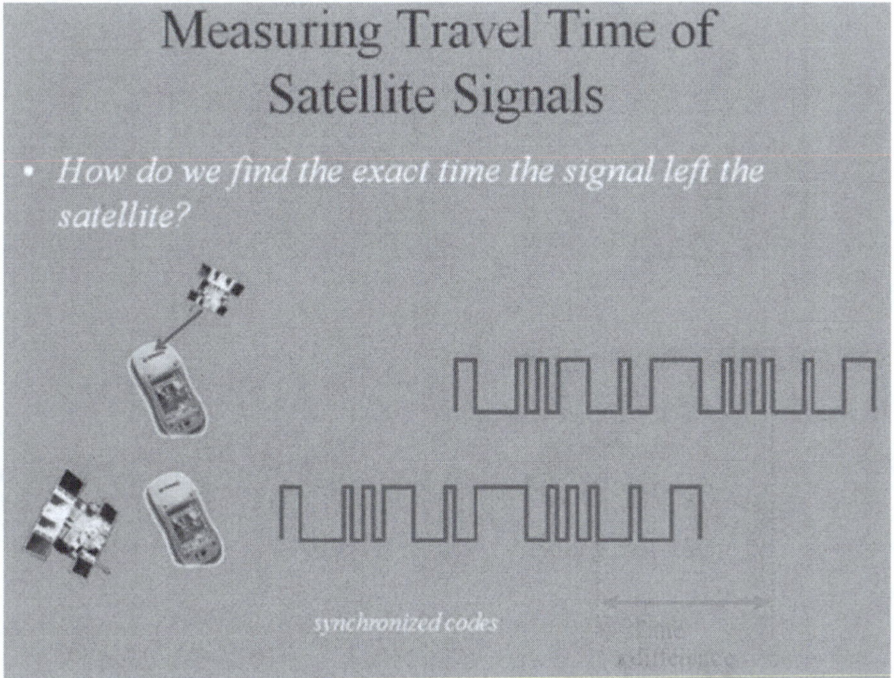

Figure 4.2 Measuring GPS (See colour version on page 172)

Using the Distance Measurements to Calculate a Position

Once the receiver has the distance measurements, it's basically a problem of geometry. If it "knows" where the four satellites are, and how far it is from each satellite, it can compute its location through trilateration. Here's an illustration of how it works.

1. The GPS receiver "locks on" to one satellite and calculates the range to be 12,000 miles. This fact helps narrow the receiver location down, but it only tells us that we are somewhere on a sphere which is centered on the satellite and has a 12,000 mile radius. Many of the locations on that sphere are not on earth, but out in space.

2. Now, consider that the receiver picks up a signal from a second satellite and calculates the range between the receiver and the satellite to be 11,000 miles. That means we are also somewhere on a sphere with an 11,000 mile radius with the second satellite at the center. We must, therefore, be somewhere where these two spheres

intersect. When the two spheres intersect, a circle is formed, so we must be somewhere on that circle.

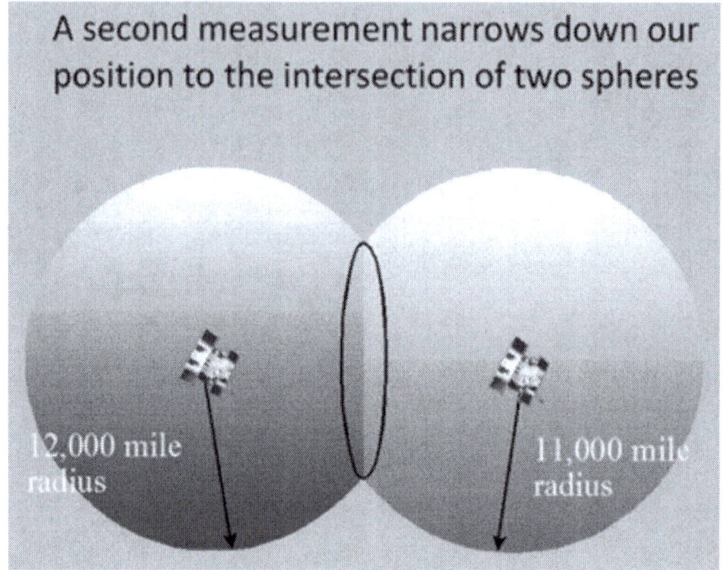

Figure 4.3 GPS Measurement

3. If the receiver picks up another satellite, say at 11,500 miles away, another sphere is formed, and there are only two points where the three spheres intersect.

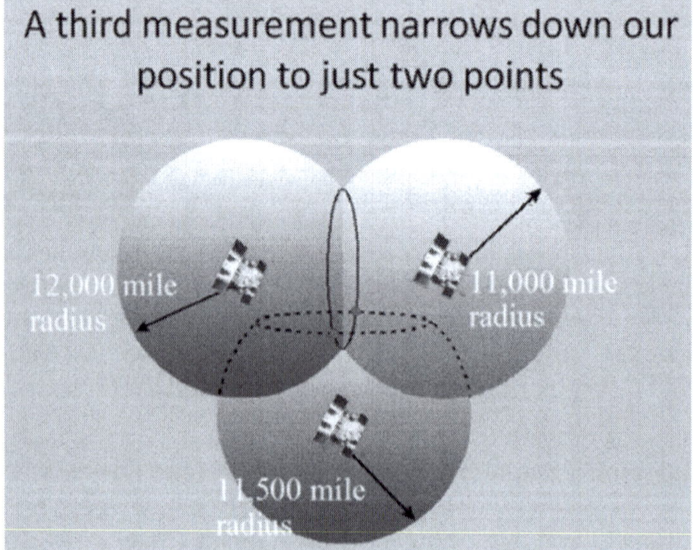

Figure 4.4 GPS Measurement own position

Usually the receiver can discard one of the last two points because it is nowhere near the earth. So, we're left with one point which is the location of the GPS receiver.

4. In practice, a fourth measurement is needed to correct for clock error.

GPS Error

There are many sources of possible errors that will degrade the accuracy of positions computed by a GPS receiver. The travel time of GPS satellite signals can be altered by atmospheric effects; when a GPS signal passes through the ionosphere and troposphere it is refracted, causing the speed of the signal to be different from the speed of a GPS signal in space. Sunspot activity also causes interference with GPS signals. Another source of error is measurement noise, or distortion of the signal caused by electrical interference or errors inherent in the GPS receiver itself. Errors in the ephemeris data (the information about satellite orbits) will also cause errors in computed positions, because the satellites weren't really where the GPS receiver "thought" they were (based on the information it received) when it computed the positions. Small variations in the atomic clocks (clock drift) on board the satellites can translate to large position errors; a clock error of 1 nanosecond translates to 1 foot or 3 meters user error on the ground. Multipath effects arise when signals transmitted from the satellites bounce off a reflective surface before getting to the receiver antenna. When this happens, the receiver gets the signal in straight line path as well as delayed path (multiple paths). The effect is similar to a ghost or double image on a TV set.

Geometric Dilution of Precision (GDOP)

Satellite geometry can also affect the accuracy of GPS positioning. This effect is called Geometric Dilution of Precision (GDOP). GDOP refers to where the satellites are in relation to one another, and is a measure of the quality of the satellite configuration. It can magnify or lessen other GPS errors. In general, the wider the angle between satellites, the better the measurement *(see GPS Basics slide show for an illustration)*. Most GPS receivers select the satellite constellation that will give the least uncertainty, the best satellite geometry.

GPS receivers usually report the quality of satellite geometry in terms of Position Dilution of Precision, or PDOP. PDOP refers to horizontal (HDOP) and vertical (VDOP) measurements (latitude, longitude and altitude). You can check the quality of the satellite configuration your receiver is currently using by looking at the PDOP value. A low DOP indicates a higher probability of accuracy, and a high DOP indicates a lower probability of accuracy. A PDOP of 4 or less is excellent, a PDOP between 5 AND 8 is acceptable, and a PDOP of 9 or greater is poor. Another term you may encounter is TDOP, or Time Dilution of Precision. TDOP refers to satellite clock offset. On a GPS receiver you can set a parameter known as the PDOP mask. This will cause the receiver to ignore satellite configurations that have a PDOP higher than the limit you specify.

Selective Availability (SA)

Selective Availability, or SA, occurred when the DoD intentionally degraded the accuracy of GPS signals by introducing artificial clock and ephemeris errors. When SA was implemented, it was the largest component of GPS error, causing error of up to 100 meters. SA is a component of the Standard Positioning Service (SPS), which was formally implemented on March 25, 1990, and was intended to protect national defense. SA was turned off on May 1, 2000.

Lists the possible sources of GPS error and their general impact on positioning accuracy.

Table 4.1. GPS Error Budget

Error source	Potential error	Typical error
Ionosphere	5.0 meters	0.4 meters
Troposphere	0.5 meters	0.2 meters
Ephemeris data	2.5 meters	0 meters
Satellite clock drift	1.5 meters	0 meters
Multipath	0.6 meters	0.6 meters
Measurement noise	0.3 meters	0.3 meters
Total	~ 15 meters	~ 10 meters

How to Reduce GPS Error

You've probably heard people talk about getting 1 to 5 meter accuracy with a GPS receiver, or even centimeter or millimeter accuracy. Is there a way to cancel out the errors and get better than 15 meter accuracy? The answer is yes, but the level of accuracy depends on the type of equipment you are using. The following discussion describes a technique used to achieve 1 to 5 meter accuracy using mapping (resource) grade receivers. Some mapping grade receivers are even capable of sub-meter accuracy, but the increased accuracy comes at a price. Survey grade receivers are the most accurate, capable of centimeter or even millimeter accuracy, depending on the equipment, but they use more advanced techniques to achieve this level of accuracy and, naturally, are more expensive. Recreational grade receivers usually can receive real-time differential corrections, but they cannot store a file that can be differentially corrected using post-processing methods.

Differential Correction

Differential correction is a method used to reduce the effects of atmospheric error and other sources of GPS positioning error (differential correction cannot correct for multipath or receiver error; it counteracts only the errors that are common to both reference and roving receivers). It requires, in addition to your "roving" GPS receiver, a GPS receiver on the ground in a known location to act as a static reference point. This type of setup is often called a GPS base station. Since the base station "knows" where it is, it can compute the errors in its position calculations (in reality, it computes timing errors) and apply them to any number of roving receivers in the same general area. This requires that the base and rover receivers "see" the same set of satellites at the same time.

The base station, depending upon how it is configured, can correct roving GPS receiver data in one (or both) of two ways. 1) In the first method, called real-time differential correction or real-time differential GPS (DGPS), the base station transmits (usually via radio link) error correction messages to other GPS receivers in the local area. In this case, the positions you read on your GPS receiver while you are out collecting data, are the corrected positions. 2) The second method, called post-processed differential correction, is performed on a computer *after* the roving receiver data are collected. While you are out in the field collecting data, the positions you read on your roving GPS receiver are uncorrected. It is not until you take your rover files back to the office and process them using differential correction software and data from the base station file, that you get corrected positions. The base station file contains information about the timing errors. This information allows the differential correction software to apply error corrections to the roving receiver file during processing. Since the base and rover receivers have to "see" the same set of satellites at the same time, the base file has to start before the rover file starts, and end after the rover file ends (a base station is normally set up to track all satellites in view, insuring that it will "see" at least the four satellites that the roving receiver is using to compute positions). Post-processed differential correction, then, requires both base and rover receivers that are capable of collecting and storing files. Most recreational grade receivers cannot collect and store files that can be differentially corrected.

Differential Correction Sources

Several options are available for obtaining differential corrections: 1) Use a local base station, 2) Use one of the wide-area differential GPS (WADGPS) services.

You can set up your own local base station or share a base station with other GPS users in your area. If you are using post-processed differential correction, the base station can usually serve users in an area with about a 2 to 300 mile radius (the further the roving receiver is from the base station, the less accurate the corrections). If you are using real-time differential correction, you must establish radio links, and your coverage area is limited by the strength of the radio transmissions. If you plan to set up your own base station, make sure the manufacturer can supply all the necessary components including base and rover receivers, radios (if using real-time correction) and differential correction software (if using post-processed correction).

Many government agencies operate GPS base stations and may provide correction files for post-processed differential correction. If you plan to use files from an operating base station, determine the manufacturer of the base station receiver. If you purchase a roving receiver from the same manufacturer, the base and rover files will be compatible and your rover files can be differentially corrected using software provided by the manufacturer. If your rover is made by a different manufacturer, you will probably have to convert the base files to Receiver Independent Exchange (RINEX) format before they can be used to differentially correct your rover data. Make sure your differential correction software can use a RINEX base file. If not, the rover file has to be converted to RINEX format and then differentially corrected using software provided by the base station manufacturer. In this situation, any attribute data stored in your roving receiver file will be lost because the RINEX format supports conversion of position data only. If you need to use a RINEX conversion, make sure you test it thoroughly before purchasing a receiver.

Some companies, such as Omnistar and RACAL provide differential corrections in real-time via their own communication satellite systems. To receive their signals you must purchase a special satellite receiver as well as the subscription service. The signals from satellites are generally available over a widespread area, hence the term wide-area differential GPS (WADGPS). Your GPS receiver must be able to receive the correction data from the satellite receiver and apply those corrections to the data it collects. Some companies offer an integrated GPS/satellite correction receiver so you don't have to purchase a separate GPS receiver. Just be sure the system will allow attribute data collection and can provide any other features you need.

The Nationwide Differential GPS (NDGPS) system is another source for differential correction data. NDGPS stations exist around the country, and the system is currently being expanded, with the hope of providing full coverage throughout the continental United States. An additional source of real-time differential correction data is the Wide Area Augmentation System (WAAS) operated by the Federal Aviation Administration (FAA). Many receivers are now WAAS compatible.

❑❑❑

5

Introduction to Geographical Information System

Introduction

A **geographic information system** or **geographical information system** (**GIS**) is a system designed to capture, store, manipulate, analyze, manage, and present all types of spatial or geographical data. The acronym GIS is sometimes used for geographic information science (GIScience) to refer to the academic discipline that studies geographic information systems and is a large domain within the broader academic discipline of geoinformatics. What goes beyond a GIS is a spatial data infrastructure, a concept that has no such restrictive boundaries. In a general sense, the term describes any information system that integrates stores, edits, analyzes, shares, and displays geographic information. GIS applications are tools that allow users to create interactive queries (user-created searches), analyze spatial information, edit data in maps, and present the results of all these operations. Geographic information science is the science underlying geographic concepts, applications, and systems.

GIS is a broad term that can refer to a number of different technologies, processes, and methods. It is attached to many operations and has many applications related to engineering, planning, management, transport/logistics, insurance, telecommunications, and business. For that reason, GIS and location intelligence applications can be the foundation for many location-enabled services that rely on analysis and visualization. GIS can relate unrelated information by using location as the key index variable. Locations or extents in the Earth space–time may be recorded as dates/times of occurrence, and x, y, and z coordinates representing, longitude, latitude, and elevation, respectively. All Earth-based spatial-temporal location and extent references should, ideally, be relatable to one another and ultimately to a "real" physical location or extent. This key characteristic of GIS has begun to open new avenues of scientific inquiry.

Components of a GIS

A working GIS integrates five key components: hardware, software, data, people, and methods.

Hardware

Hardware is the computer on which a GIS operates. Today, GIS software runs on a wide range of hardware types, from centralized computer servers to desktop computers used in stand-alone or networked configurations.

Software

GIS software provides the functions and tools needed to store, analyze, and display geographic information. Key software components are

- Tools for the input and manipulation of geographic information
- A database management system (DBMS)
- Tools that support geographic query, analysis, and visualization
- A graphical user interface (GUI) for easy access to tools.

Data

Possibly the most important component of a GIS is the data. Geographic data and related tabular data can be collected in-house or purchased from a commercial data provider. A GIS will integrate spatial data with other data resources and can even use a DBMS, used by most organizations to organize and maintain their data, to manage spatial data.

People

GIS technology is of limited value without the people who manage the system and develop plans for applying it to real-world problems. GIS users range from technical specialists who design and maintain the system to those who use it to help them perform their everyday work.

Methods

A successful GIS operates according to a well-designed plan and business rules, which are the models and operating practices unique to each organization.

Definitions of GIS

GIS is a special-purpose digital database in which a common spatial coordinate system is the primary means of reference. Comprehensive GIS require a means of:

1. Data input, from maps, aerial photos, satellites, surveys, and other sources
2. Data storage, retrieval, and query
3. Data transformation, analysis, and modeling, including spatial statistics
4. Data reporting, such as maps, reports, and plans.

Three observations should be made about this definition:

First, GIS are related to other database applications, but with an important difference. All information in a GIS is linked to a spatial reference. Other databases may contain locational information (such as street addresses, or zip codes), but a GIS database uses geo-references as the primary means of storing and accessing information.

Second, GIS integrates technology. Whereas other technologies might be used only to analyze aerial photographs and satellite images, to create statistical models, or to draft maps, these capabilities are all offered together within a comprehensive GIS.

Third, GIS, with its array of functions, should be viewed as a process rather than as merely software or hardware. GIS are for making decisions. The way in which data is entered, stored, and analyzed within a GIS must mirror the way information will be used for a specific research or decision-making task. To see GIS as merely a software or hardware system is to miss the crucial role it can play in a comprehensive decision-making process.

Geographic information systems are among the most exciting and powerful Geomatics decision-making tools in the world. A GIS uses computer technology to integrate, manipulate and display a wide range of information to create a picture of an area's geography, environment and socio-economic characteristics. Beginning with a computerized topographic map as its base, a GIS *overlays and integrates* graphic and textual information from separate databases. The end result is a customized and reliable tool that can support decision making and problem solving and provide almost instantaneous answers to complex questions. (Geomatics Canada Web Site, 2000).

In their book *"Geographic Information Systems: An Introduction"* published in 1990, Star and Estes define GIS as an information system designed to work with data referenced by spatial or geographic coordinates. According to the same authors, a GIS is both a database system with specific capabilities for spatially-referenced data, as well as a set of operations for working with the data, i.e., analyzing it.

Lang (2000) provides this definition: "A geographic information system (GIS) is a computer system for analyzing and mapping just about anything, moving or stationary. A GIS integrates common database operations, such as query and statistical analysis, with the ability to see how data relates in space and time. The maps produced with a GIS are useful for showing places and the events that occur there, like outbreaks of disease. They are useful for analyzing and visualizing any system that's spatial or spatio-temporal (that also changes with time), for mapping a patient's heart or brain, for instance, or showing a breakdown of diagnoses on a map of the body, or even indicating which beds on a hospital floor are occupied, for how long, and by whom."

GIS is not merely a software package. William Henriques defines GIS as an organized collection of computer hardware, software, geographic/spatial data, complementary technologies, and personnel designed to efficiently capture, store, update, manipulate, analyze, and display all forms of geographically referenced information. The user becomes part of the GIS whenever complicated analyses, such a *spatial analysis* and modeling, are carried out.

Hall (1999) stresses the point that GIS are much more than simply computer mapping systems and can do much more than merely outputting improved digital versions of static, paper-based maps. GIS blend map production, image presentation and statistical analysis capabilities into a powerful analytical tool that can be applied to a variety of problems.

Another way to describe GIS would be as information management tools for producing "*smart maps*" where the geographic features on these maps are linked to tables of information about them. A spreadsheet or database table alone is not easy to interpret (even with the help of conventional database and spreadsheet manipulation and querying tools). These GIS "smart maps" go beyond conventional spreadsheet and database tables, helping us discover and visualise new patterns and relationships by allowing different information layers to be matched, interlinked, queried and analysed, thus producing new knowledge and hypotheses for further investigation. This can be considered a form of data mining.

A database table like this one is very difficult to interpret and many very useful patterns in it might remain hidden and undiscovered.

Today, geographic information systems are commonly used for everything one can imagine, from basic mapping to supporting resource exploration and development, from environmental management to the planning and administration of transportation and telecommunications systems, utility infrastructures, urban development and land use.

GIS Concepts

To use GIS effectively, it is important to understand the basic GIS terminology and functionality. While each GIS software has slightly different naming conventions, there are certain principles common to all systems. At first, we briefly describe the GIS basics in general (for in depth information read Longley et al., 2005, Clarke, 2002, or Burrough and McDonnell, 1998) and then we explain the principles of map projections and coordinate systems that are used to Georeference the data.

Data, Dataset and Database

Data is information represented in the format of digit, letter and symbol used to describe status, behavior and their consequence of geographical object. There are some inner relations and different between data and information, as defined above, data indicates those value recorded and stored in computer, the meaning of the value represented is information. *Dataset* is the minimum body of data used for data transform, storage, manipulation, copying, and other activities. Usually, there is one type of spatial data feature as point, line or polygon employed to represent one kind of geographical object such as river or topography or building. In most cases, data layer have the same meaning with dataset, but a few data layer can be organized into one dataset in some special occasions.

Database, as the word per se means data and base, is the combination of dataset according to the defined logical principles. Usually, the dataset in one database share the same data structure, data storage method, data format and similar data management interface. Except the dataset contained, database itself has some functions as data updating, data manipulation (extracting, clipping, overlaying, statistics), and user propriety definition. For geo-spatial database, database management system is often one part of GIS software package or some other DBMS database connected to GIS software package with some inter software component such as Arc SDE of ESRI. Figure 3.2 shows the concept idea of the Nam Chuen Watershed Database implemented by the Natural Resources and Environmental Management Project in Thailand (NREM, 2000).

Spatial Elements

Spatial objects in the real world can be thought of as occurring as four easily identifiable types: *Points, Lines, Areas,* and *Surfaces* (Figure 3.4). Collectively, they can represent most of the tangible natural and human phenomena that we encounter on an everyday basis. In general, points, lines, and areas are used to explicitly represent real-world objects, where as surfaces are mostly used for volumetric representation, such as to represent hills, valleys. Thus, all data can be considered to be explicitly spatial.

Point features are spatial phenomena each of which occurs at one location in space. Each feature is said to be discrete in that it can occupy only a given point in space at any time and considered to have *no spatial dimension* - no width or length. Example of such feature would be a house or a village. But a village can be represented by point feature or area feature as well depending upon the resolution of data.

Line features are conceptualized as occupying only a *single dimension* in coordinate space. They are represented as the series of single coordinates connected to each other. Roads, rivers, are the examples of linear features. The resolution or scale of given dataset once again places a fundamental limitation to conceive them as having any width. Linear features, unlike point features, allow us to measure their spatial extent/ length. *Area features* have *two dimensions* both length and width dimensions. Area is composed of series of lines that begin and end at the same location. We can describe their shapes and orientations, and the amount of territory occupied as well. In database, the term polygon is often used instead of area. Again, physical size in relation to the scale determines whether an object is represented by an area or by a point.

It is often that area is divided into regular squares or rectangles so that all objects are described in terms of areas. This entire data structure is called a *grid*. Each square or rectangular is known as a cell and represents a uniform value.

Adding the dimension of height to area features allows us to observe and record the existence of *Surfaces.* Surfaces have three dimensions length, width, and height. For instance, hills, valleys, and ridges can be described by citing their locations, amount of area they occupy, how they are oriented, and by noting their heights.

Spatial Measurement Levels

All these spatial feature or entities or objects we observe contain information not only about how they occupy space but also about what they are and how important they to what we are studying. For example a tree, viewed as a point in the landscape, might be classified based on age class: seedling, sapling and tree. The additional non-spatial information that helps us describe the objects we observe in space comprises the feature's attributes. However before we assign these properties or attributes, we must know how to measure them. There are four levels of measurement in terms of commonly used geographic features. The first two are the *nominal* scale and the *ordinal* scale, in which observations are simply classified mutually exclusive categories. The final two scales, the *interval* and *ratio*, are those we ordinarily think of as "*measurements*" because they involve determination of a classification of observation (Davis, 1986).

Nominal scale - is based upon classification of data into mutually exclusive groups or categories. This falls in *qualitative* category of attribute data. These categories may be identified by names, such as "red", "green" or "blue", as well as by types for examples "swamp", "desert" or "forest". Number may be used simply as identifiers, for instance village code. These can be no connotation that "2" is twice as much as "1" or "5" is greater than "4". Therefore, no attempt is made to quantify the size of the groups.

Ordinal scale – Allows to rank the classes in a hierarchy of states, but not to distinguish between them on the basis of the size of a measured variable. For example, the minerals on the scale, which extends from one to ten, increase in hardness with higher rank, the successive states are not equal.

Interval scale – Interval scale of measurement is used when we have equal intervals between the measurement units. However, an interval scale has *no natural zero or no true starting point*. Such data can be compared, as in the case of ordinal data but yet more precise. A good example of spatial data consists of soil temperatures across a study area containing widely different soil types. Say for example, the soil types measured at the same time exhibit a difference of 400 F (i.e. 800 F for dark and 400 F for light soil). But one limitation remains, i.e. can we say the dark soil is twice as warm as the light soil? This is not the case, when we convert these values to Kelvin scale.

Ratio scale – Ratio scale is the highest form of measurement. It not only has interval scale between steps, but also has a *true zero point*.

Measurements of length or conversion of Fahrenheit temperatures to Kelvin scale are of this type. Therefore, ratio scale allows us to make a direct comparison between two spatial variables. For example, a tree ten meters high is twice the height of a tree 5 meters high. And a tree is zero height does not exist because it has no length at all.

Data Sources

What is Special about Spatial Data?

An attention to location, spatial interaction, spatial structure and spatial processes lies at the heart of research in several sub disciplines in the social sciences. Empirical studies in these fields routinely employ data for which locational attributes (the "where") are an important source of information. Such data typically consist of one or a few cross-sections of observations for either.

Micro-units, such as households, store sites, settlements, or for aggregate spatial units, such as electoral districts, counties, states or even countries. Observations such as these, for which the absolute location and/or relative positioning (spatial arrangement) is taken into account are referred to as spatial data. In the social sciences, they have been utilized in a wide range of studies, such as archeological investigations of ancient settlement patterns (e.g., in Whitley and Clark, 1985, and Kvamme, 1990), sociological and anthropological studies of social networks (e.g., in White et al., 1981, and Doreian et al., 1984), demographic analyses of geographical trends in mortality and fertility (e.g., in Cook and Pocock, 1983, and Loftin and Ward, 1983), and political models of spatial patterns in international conflict and cooperation (e.g., in O'Loughlin, 1985, and O'Loughlin and Anselin, 1991). Furthermore, in urban and regional economics and regional science, spatial data are at the core of the field and are studied to model the spatial structure for a range of socioeconomic variables, such as unemployment rates (Bronars and Jansen, 1987), household consumer demand (Case, 1991), and prices for gasoline (Haining, 1984) or housing (Dubin, 1992).

The locational attributes of spatial data (i.e., for the settlements, households, regions, etc.) are formally expressed by means of the geometric features of points, lines or areal units (polygons) in a plane, or, less frequently, on a surface. This spatial referencing of observations is also the salient feature of a Geographic Information System (GIS), which make it a natural tool to aid in the analysis of spatial data. I return to this issue in more detail below.

The crucial role of location for spatial data, both in an absolute sense (coordinates) and in a relative sense (spatial arrangement, distance) has major implications for the way in which they should be treated in statistical analysis, as discussed in detail in Anselin (1990a). Indeed, location gives rise to two classes of socalled spatial effects: spatial dependence and spatial heterogeneity. The first, often also referred to as spatial autocorrelation or spatial association, follows directly from Tobler's (1979) First Law of Geography, according to which "everything is related to everything else, but near things are more related than distant things." As a consequence, similar values for a variable will tend to occur in nearby locations, leading to spatial clusters. For example, a high crime neighbourhood in an inner city will often be surrounded by other high crime areas, or a low income county in a remote region may be neighbouring other low income counties. This spatial clustering implies that many samples of geographical data will no longer satisfy the usual statistical assumption of independence of observations.

A major consequence of the dependence in a spatial sample is that statistical inference will not be as efficient as for an independent sample of the same size. In other words, the dependence leads to a loss of information.1 Roughly speaking, and everything else being the same, this will be reflected in larger, variances for estimates, lower significance levels in tests of hypotheses and a poorer fit for models estimated with data from dependent samples, compared to independent samples of the same size. I will refer to this aspect of spatial dependence in the rest of the paper as a nuisance. The loss in efficiency may be remedied by increasing the sample size or by designing a sampling scheme that spaces observations such that their interaction is negligible. Alternatively, it may be taken into account by means of specialized statistical methods. In this paper, I will focus on the latter. When spatial dependence is considered to be a nuisance, one only wants to make sure that the interpretation of the results of a statistical analysis are valid. One is thus not really interested in the source of the spatial association, i.e., in the form of the spatial interaction, the characteristics of the spatial structure, or the shape of the spatial and/or social processes that led to the dependence. When the latter is the main concern, I will use the term substantive spatial dependence instead.

The second type of spatial effect, spatial heterogeneity, pertains to the spatial or regional differentiation which follows from the intrinsic uniqueness of each location. This is a special case of the general problem of structural instability. As is well known, in order to draw conclusions with a degree of general validity from the study of a spatial sample, it is necessary that this sample represents some type of equilibrium. In the analysis of cross-sectional data in the social sciences this assumption is typically made. However, this assumption is considered with respect to the time dimension only, and systematic instability or structural variation that may be exhibited across different locations in space is mostly ignored. Such spatial heterogeneity may be evidenced in various aspects.

The History of Spatial Analysis

Spatial analysis can perhaps be considered to have arisen with the early attempts at cartography and surveying but many fields have contributed to its rise in modern form. Biology contributed through botanical studies of global plant distributions and local plant locations, ethological studies of animal movement, landscape ecological studies of vegetation blocks, ecological studies of spatial population dynamics, and the study of biogeography. Epidemiology contributed with early work on disease mapping, notably John Snow's work mapping an outbreak of cholera, with research on mapping the spread of disease and with locational studies for health care delivery. Statistics has contributed greatly through work in spatial statistics. Economics has contributed notably through spatial econometrics. Geographic information system is currently a major contributor due to the importance of geographic software in the modern analytic toolbox. Remote sensing has contributed extensively in morphometric and clustering analysis. Computer science has contributed extensively through the study of algorithms, notably in computational geometry. Mathematics continues to provide the fundamental tools for analysis and to reveal the complexity of the spatial realm, for example, with recent work on fractals and scale invariance. Scientific modeling provides a useful framework for new approaches.

The Concept of Spatial Data

Spatial Data

The term "spatial data" refers to the representation of multi-dimensional data, such as points, lines and surfaces, as a list of numbers using a particular coordinate system. A typical example of spatial data is electronic map data, which is used to represent the topography of the real world in a coordinate system.

There are two fundamentally different kinds of spatial data: raster data and vector data.

In a vector model, points define coordinates, and points and lines define the borders between different aspects of the real world. The location of each of these aspects on a map is specified and maintained using a consistent coordinate system. Points, lines and polygons are used to represent coordinates or geographical features that are irregularly distributed in the real world. Lines are used to represent one-dimensional features such as roads, and polygons are used to represent two-dimensional features such as forests and the like is defined as a matrix of the pixels and cells in which the aspect or set of coordinates exists. The level of detail that it is possible to represent using a raster model depends on the cell size. The area in each cell cannot be divided any further; that is, all of the attributes that apply to the cell apply uniformly to the entire area within the cell. All cells are identical in size. A raster model typically comprises millions of cells.

The basic units in a vector model are points, lines and polygons. Compared to a raster model, a vector model typically comprises far fewer basic units, and moreover they are not uniform in size. A vector file typically contains thousands of elements, the specific location of each element being defined by one or more consecutive coordinate values. Vector data are easier to manipulate on a computer than raster data because the number of data items is lower, and additionally because the model is easier to adjust for different scales, and yields more precise results when the scale is changed.

Fundamental Issues in Spatial Analysis

Spatial analysis confronts many fundamental issues in the definition of its objects of study, in the construction of the analytic operations to be used, in the use of computers for analysis, in the limitations and particularities of the analyses which are known, and in the presentation of analytic results. Many of these issues are active subjects of modern research.

Common errors often arise in spatial analysis, some due to the mathematics of space, some due to the particular ways data are presented spatially, some due to the tools which are available. Census data, because it protects individual privacy by aggregating data into local units, raises a number of statistical issues. The fractal nature of coastline makes precise measurements of its length difficult if not impossible. Computer software fitting straight lines to the curve of a coastline, can easily calculate the lengths of the lines which it defines. However these straight lines may have no inherent meaning in the real world, as was shown for the coastline of Britain.

These problems represent one of the greatest dangers in spatial analysis because of the inherent power of maps as media of presentation.

When results are presented as maps, the presentation combines the spatial data which is generally very accurate with analytic results which may be grossly inaccurate. Some of these issues are discussed at length in the book *How to Lie with Maps.*

Spatial characterization

Spread of bubonic plague in medieval Europe. The colors indicate the spatial distribution of plague outbreaks over time. Possibly due to the limitations of printing or for a host of other reasons, the cartographer selected a discrete number of colors to characterize (and simplify) reality.

The definition of the spatial presence of an entity constrains the possible analysis which can be applied to that entity and influences the final conclusions that can be reached. While this property is fundamentally true of all analysis, it is particularly important in spatial analysis because the tools to define and study entities favor specific characterizations of the entities being studied. Statistical techniques favor the spatial definition of objects as points because there are very few statistical techniques which operate directly on line, area, or volume elements. Computer tools favor the spatial definition of objects as homogeneous and separate elements because of the limited number of database elements and computational structures available, and the ease with which these primitive structures can be created.

Data Model

Types of Database Management Systems

A DBMS can take any one of the several approaches to manage data. Each approach constitutes a database model. A data model is a collection of descriptions of data structures and their contained fields, together with the operations or functions that manipulate them. A data model is a comprehensive scheme for describing how data is to be represented for manipulation by humans or computer programs. A thorough representation details the types of data, the topological arrangements of data, spatial and temporal maps onto which data can be projected, and the operations and structures that can be invoked to handle data and its maps. The various ***Database Models are the following:-***

- Relational - data model based on tables.
- Network - data model based on graphs with records as nodes and relationships between records as edges.
- Hierarchical - data model based on trees.

- Object-Oriented - data model based on the object-oriented programming paradigm.

Relational Model

The basic building block of the relational model is the table, which is a matrix of rows and columns. Tables are related to each other via a common entity characteristic or attribute (primary key in the parent table is a foreign key in the child table). The parent table is the table which maps to the entity of the "1" side of the relationship and the child table maps to the entity of the "many" side of the relationship between the two tables.

All three relationship types are easily represented in this model. One of the disadvantages of the relational model is that it requires substantial system overhead to run the Relational DBMS (RDBMS). However, with the currently available advanced computer hardware and software, high requirements for processing relational databases do not represent an overhead problem any more.

Basic Structure

The relational data model is implemented through a very sophisticated relational database management system (RDBMS). The RDBMS performs the same basic functions provided by the hierarchical and network database systems, plus a host of other functions that make the relational data model easier to understand and to implement.

The most important advantage of the RDBMS is its ability to let the user/designer operate in a human logical environment. The RDBMS manages all of the complex physical details. Thus, the relational database is perceived by the user to be a collection of tables in which data are stored.

Each table is a matrix consisting of series of row/column intersections. Tables, also called relations, are related to each other by sharing a common entity characteristic/attribute. For example, the CUSTOMER table in Figure 5.1 might contain a sales agent's number which maintains a common link to the agent table.

The common link between the CUSTOMER and AGENT tables thus enables us to match the customer to his/her sales agent, even though the customer data are stored in another table. Although the tables are completely independent of one another, we can easily connect the data between tables. The relational model thus provides a minimum level of controlled redundancy to eliminate most of the redundancies found in old file systems.

Database name: Ch02_InsureCo Table name: AGENT (first six attributes)

AGENT_CODE	AGENT_LNAME	AGENT_FNAME	AGENT_INITIAL	AGENT_AREACODE	AGENT_PHONE
501	Alby	Alex	B	713	228-1249
502	Hahn	Leah	F	615	882-1244
503	Okon	John	T	615	123-5589

Link through AGENT_CODE

Table name: CUSTOMER

CUS_CODE	CUS_LNAME	CUS_FNAME	CUS_INITIAL	CUS_AREACODE	CUS_PHONE	CUS_RENEW_DATE	AGENT_CODE
10010	Ramas	Alfred	A	615	844-2573	05-Apr-2004	502
10011	Dunne	Leona	K	713	894-1238	16-Jun-2004	501
10012	Smith	Kathy	W	615	894-2285	29-Jan-2005	502
10013	Olowski	Paul	F	615	894-2180	14-Oct-2004	502
10014	Orlando	Myron		615	222-1672	28-Dec-2004	501
10015	O'Brian	Amy	B	713	442-3381	22-Sep-2004	503
10016	Brown	James	G	615	297-1228	25-Mar-2004	502
10017	Williams	George		615	290-2556	17-Jul-2004	503
10018	Farriss	Anne	G	713	382-7185	03-Dec-2004	501
10019	Smith	Olette	K	615	297-3809	14-Mar-2004	503

Figure 5.1 Relationship Type

The relationship type (1:1, 1:M, or M:N) is often shown in a relational schema, an example of which is depicted in Figure 5.2. A relational schema is a visual representation of the relational database's entities, the attributes within those entities, and the relationship between those entities.

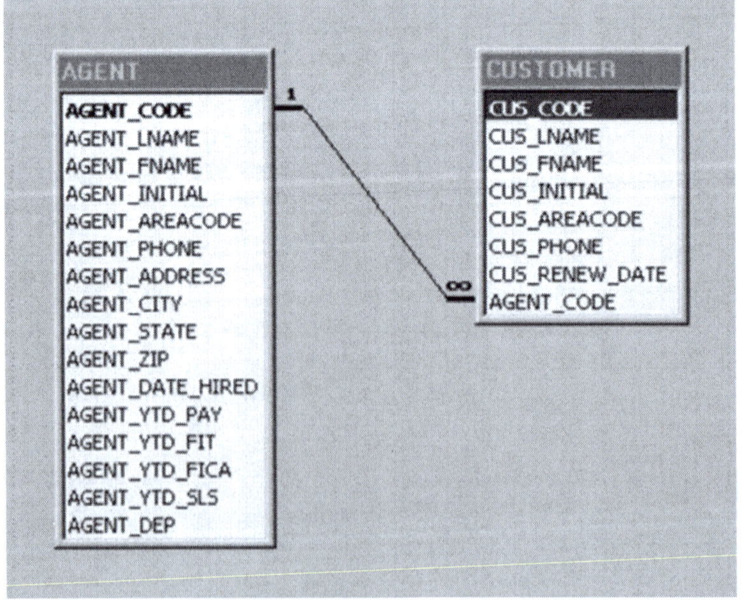

Figure 5.2 Relational schema

As you examine Figure 5.2, note that the relational scheme shows the connecting fields (in this case, AGENT_CODE) and the relationship type, 1:M. MS Access DBMS software used to generate Figure 5.2, employs the " symbol to indicate the "many" side. In this example, the CUSTOMER represents the "many" side because an AGENT can serve many CUSTOMERS. The AGENT represents the "1" side, because each CUSTOMER is served by only one AGENT.

Advantages

- Structural independence
- Improved conceptual simplicity
- Easier database design, implementation, management, and use
- Ad hoc query capability
- Powerful database management system.

Disadvantages

- Substantial hardware and system software overhead (this is not an issue any more because of the currently available hardware and software)
- Can facilitate poor design and implementation (Less experienced database designers may develop poor database design)
- May promote "islands of information" problems (because various users in different departments will be developing their own database applications).

The relational model is based on (relational) set theory in mathematics. Most commercial database products introduced in the last two decades are relational model (e.g., Oracle, Informix, DB2, Sybase, Ingres, MS SQL Server). A relation is nothing but a "table" (as perceived by us) that consists of tuples (rows) and attributes (columns).

A relation or a table has certain properties:

1. A relation consists of a set of attributes with unique names (i.e., attribute names cannot repeat in the SAME table). It doesn't matter in which order the attributes are listed.
2. The values of attributes fall into a known domain or legal values (For example, SSN consists of values from 000000000 to 999999999, or Age is between 0 and 125). These domains enforce some degree of data integrity (i.e., users don't enter meaningless values).

For example, the number of students in a class cannot be -20 or 3.26; it has to be a positive integer. Note that for a given attribute, the values should come from the same domain.

3. Each tuple (row) in the database is unique. That is, records do not repeat. To avoid records from repeating, we have what is called "Primary Key." The primary key uniquely determines each record. That is, given a value for a primary key we can uniquely determine a record. For example, SSN is the primary key for STUDENT table and there is exactly one record that matches a given student's SSN. The primary key cannot take NULL value (means a record cannot exist without a unique identifier. For example, SSN is the primary key for EMPLOYEE table. One cannot enter an employee's information without his/her SSN into the EMPLOYEE table).

Relational Keys

There are many types of keys that can be defined in the relational model. These have significant importance in maintaining data consistency and correctness in the database. We will discuss only the candidate key, primary key, alternate key, and foreign key.

Candidate Keys

A minimal set of attributes in a table that uniquely identifies a record. When there is more than one attribute in the candidate key, it is called *composite key*.

Primary Key: A candidate key that is chosen to represent a record uniquely. That is, a table may consist of many candidate keys, but ONLY ONE can be selected as a primary key. Note that a primary key can have more than one attribute. When there is more than one attribute we say that the primary key is a composite primary key. One of the common mistakes students have made is to refer to one of the attributes in the composite key as primary key. Please note that when there is more than one attribute in the primary key ALL ATTRIBUTES TOGETHER are called the primary key since they together define the record uniquely.

Alternate Keys: Any candidate key that is not selected to be a primary key can be an alternate key. Most databases allow us to define alternate keys by way of UNIQUE constraint.

Foreign Keys: An attribute or a set of attributes of one table that is matched to candidate keys of another table (or even the same table). That is, a "child" table may reference the "parent" table for appropriate attribute values. Please note that the corresponding attributes in two different tables must have the same data type and domain.

Query in GIS

Querying Data in ArcMap

Identifying, Selecting, and Finding Features

There are many ways to retrieve information about features in ArcMap™. The user can *identify* features by clicking on them in order to display their attributes. The user can *select* features by clicking on the features to highlight them and look at their records in the layer attribute table. The user can *find* features by using known information about the feature in order to search the map for that particular feature.

Identifying Features

Perhaps the fastest way to get information about a single feature is to identify it, using the Identify Tool.

To use the *Identify Tool,* the user must select it from the *Tools* Toolbar. Within the map, the user must click on the feature of interest in order to view the attribute information for that particular feature.

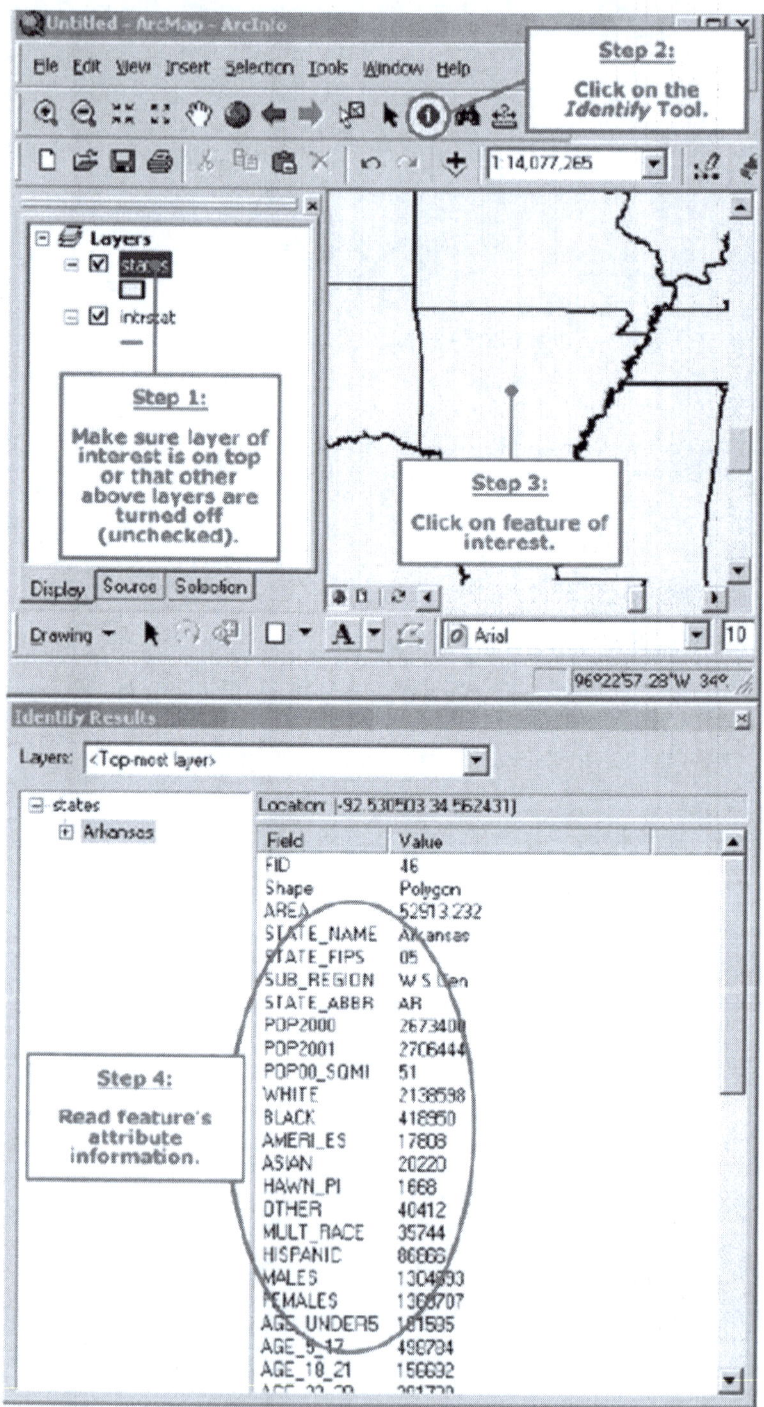

Figure 5.3 Attribute Information

Selecting Features

If the user wants to compare information about several features, the best way is to select the features on the map and look at their records in the layer attribute table. The easiest way to select multiple features is by using the ***Select Features*** Tool on the ***Tools*** Toolbar.

To use the ***Select Features*** Tool, the user must select it from the *Tools* Toolbar. On the map, all features of interest may be selected by holding down the ***shift key*** and clicking on the various features of interest. The selected features will be outlined in blue. If a feature is selected by mistake it can be deselected by holding down the ***shift key*** and clicking the feature again. All features that have been selected, can be cleared by clicking the ***Selection*** menu from the ***Standard Toolbar*** and selecting the ***Clear Selected Features*** option.

To view the selected features' attribute table, the user must rightclick on the data layer where features have been selected. The ***Open Attribute Table*** option should then be clicked.

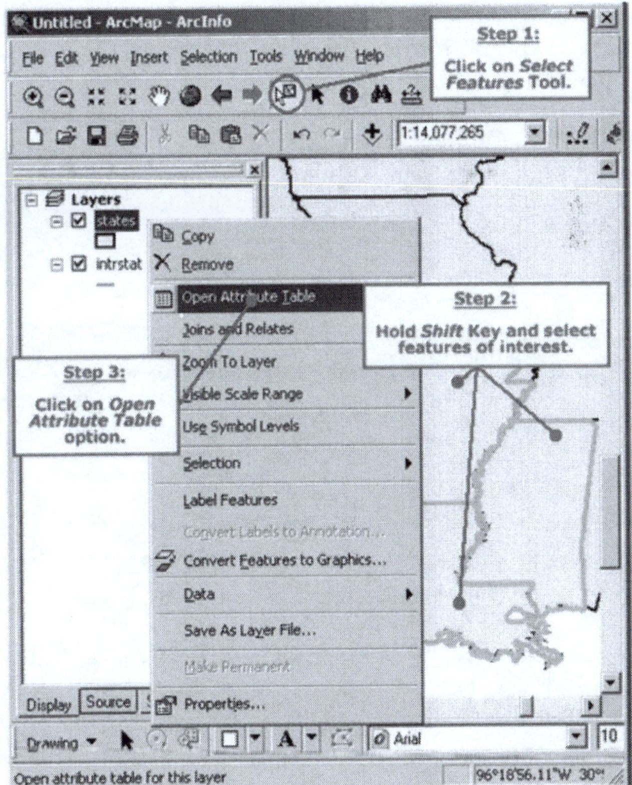

Figure 5.4 Selecting Features

After the ***Open Attribute Table*** option has been selected, the attribute table will appear with all of the selected features highlighted in blue.

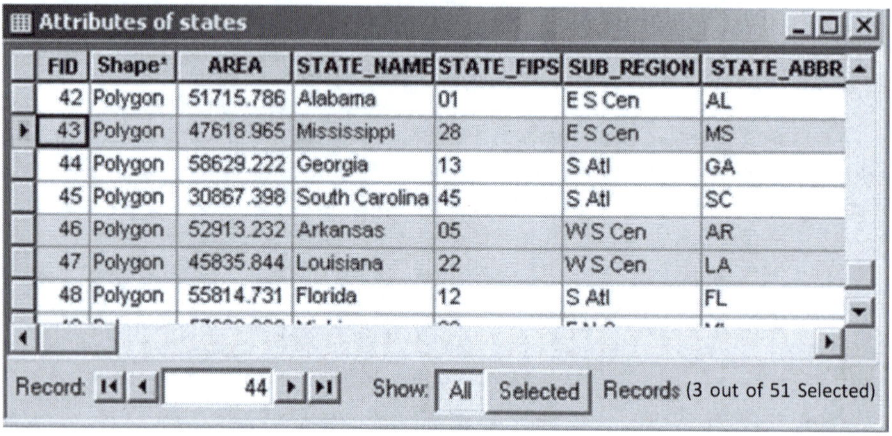

Attributes of states

FID	Shape*	AREA	STATE_NAME	STATE_FIPS	SUB_REGION	STATE_ABBR
42	Polygon	51715.786	Alabama	01	E S Cen	AL
43	Polygon	47618.965	Mississippi	28	E S Cen	MS
44	Polygon	58629.222	Georgia	13	S Atl	GA
45	Polygon	30867.398	South Carolina	45	S Atl	SC
46	Polygon	52913.232	Arkansas	05	W S Cen	AR
47	Polygon	45835.844	Louisiana	22	W S Cen	LA
48	Polygon	55814.731	Florida	12	S Atl	FL

Record: 44 Show: All Selected Records (3 out of 51 Selected)

Figure 5.5 Selected features

The user can group all of the selected attributes, by clicking on the *Selected* button at the bottom of the attribute table. Only those features that were highlighted will appear.

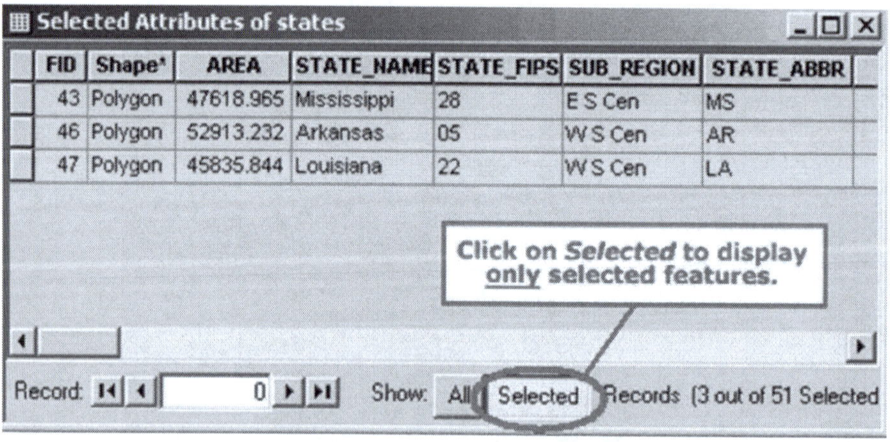

Selected Attributes of states

FID	Shape*	AREA	STATE_NAME	STATE_FIPS	SUB_REGION	STATE_ABBR
43	Polygon	47618.965	Mississippi	28	E S Cen	MS
46	Polygon	52913.232	Arkansas	05	W S Cen	AR
47	Polygon	45835.844	Louisiana	22	W S Cen	LA

Click on *Selected* to display only selected features.

Record: 0 Show: All Selected Records (3 out of 51 Selected

Figure 5.6 Selected attribute only

Now the user can easily compare various attribute values, such as Area for each state. If the user wishes to see the states listed according to largest area, the user can rightclick on the appropriate field name (in this case AREA), and the ***Sort Descending*** option may be selected from the context menu.

Finding Features

When the user has a piece of information about a feature, but is not sure where that feature is on the map, the user can search the map for that feature using the known piece of information.

The user can find a feature, by selecting the ***Find*** tool on the ***Tools*** toolbar. When the ***Find*** dialog Box appears, the ***Features*** tab should be selected. The known attribute information should be typed in the ***Find*** box. In the ***In Layers*** drop down box, the layer that the user wishes to find features in should be selected. In the ***Search*** options, the user should choose to either search all fields in the attribute table or a specific field. Once all parameters are set, the ***Find*** button should be clicked.

In the following example, the State of Arkansas was found in the U.S. State Data Layer by typing in the known attribute information (i.e. state name = Arkansas).

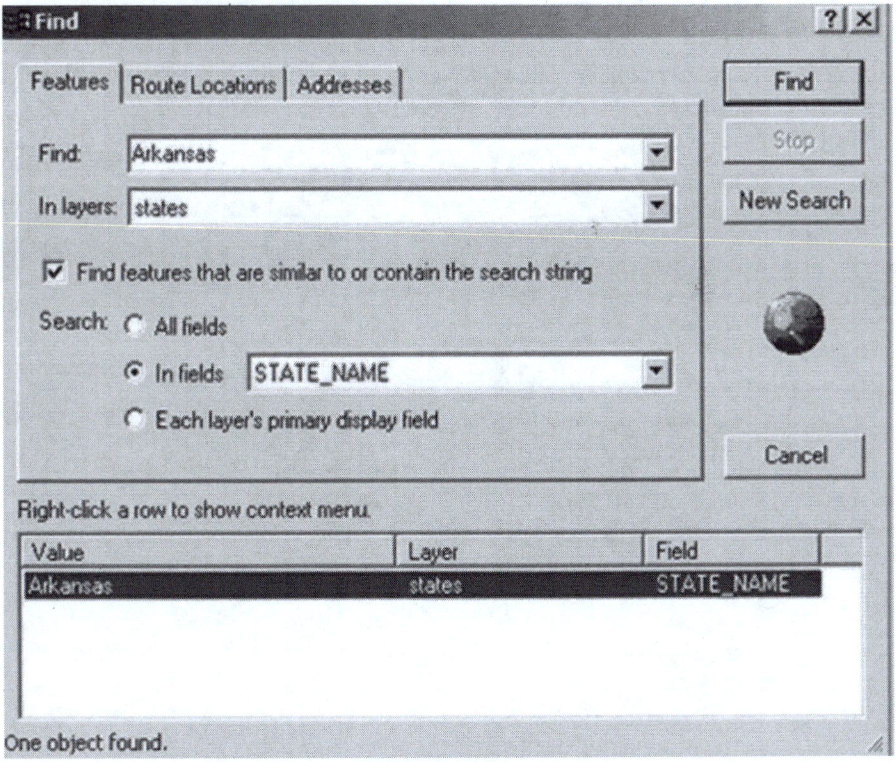

Figure 5.7 Find

Once the ***Find*** button is clicked and the feature is found, the user can locate it on the map and get its attributes. This can be done when the user right clicks on the feature row found in the ***Find*** dialog box and clicks ***Identify Feature(s)***. The feature will briefly flash within the map display and the ***Identify Results*** dialog box will open.

Selecting Features by Attributes

In addition to identifying, selecting, and finding features, the user can select features by attributes by writing a query that automatically selects features that meet a specified criteria. The simplest type of query consists of an attribute (such as STATE_NAME), a value (such as 'Arkansas'), and a relationship between the two (such as 'equal to'). A more complex query combines these simple queries using operators like 'and' / 'or'. These queries are constructed using ***Structured Query Language (SQL)***. ArcMap creates the query automatically in this format.

To create an attribute query, the user must click the ***Selection*** menu on the ***Standard*** Toolbar. The ***Select by Attributes*** option should be clicked. In the ***Select by Attributes*** dialog box, the ***Layer*** drop down arrow should be clicked and the data layer of interest should be selected.

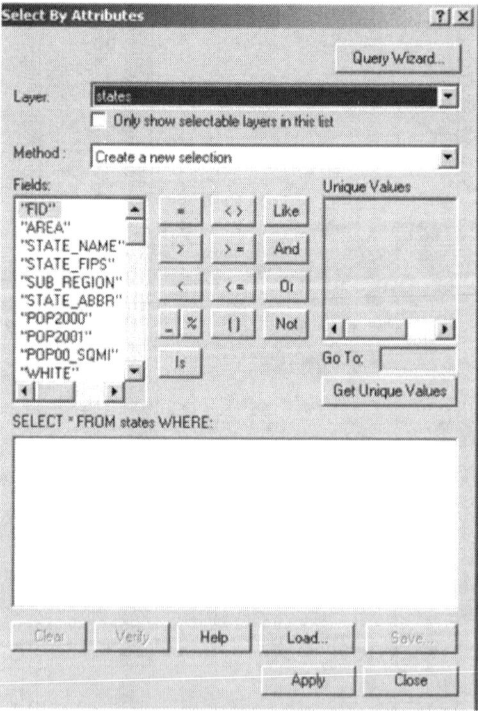

Figure 5.8 Selecting Features by Attributes

Buffering

A buffer in GIS is a zone around a map feature measured in units of distance or time. A buffer is useful for proximity analysis. A buffer is an area defined by the bounding region determined by a set of points at a specified maximum distance from all nodes along segments of an object.

Buffering in general refers the creation of a zone of a specified width around a point or a line or a polygon area. In general there are two types of buffers, namely - constant width buffers and variable width buffers. Both types can be generated for a set of coverage features based on each features attribute values. These zones or buffers can be used in queries to determine which entities occur either within or outside the defined buffer zone.

Analogous to buffering in raster GIS is distance analysis.

Buffering point data is the simplest form of buffering which involves creation of a circular polygon about each point of radius equal to the buffer distance. In this scheme the buffer distance or the radius of the circle could be fixed for all points in a layer or the user could specify it. If multiple points in the same layer are being buffered, then the buffering algorithms check for overlaps in each point's buffer and remove the overlapping sections. This process of removal of overlapping sections involves the use of intersection and dissolves.

Buffering of lines is a little more involved than buffering point data. This is complicated even more due to the fact that lines can be made up of multiple segments. Line segments are handled independently of each other. For each line segment, dx and dy between the end points are computed. Using these, two new parallel buffer lines are identified. Once these parallel buffer lines have been identified for the next segment, the line intersections of the parallel buffer lines of each segment are calculated and new coordinates are assigned to the common vertices. After that the simplest line intersection test can be used.

Lastly, the ends of the buffer are defined at the lines start and end points. Once the polygon is generated, again the overlapping sections must be removed using intersection and dissolve. Buffering of polygonal surfaces uses most of the same concepts used for line buffering. The only significant change is that the polygon buffer is created on only one side of the line that defines the polygon. Most GIS software make use of outside polygon which surrounds the area under consideration, but some make use of an inner polygon that lies inside the area under consideration.

Types of Buffer

1. Point Buffer
2. Line Buffer
3. Area or Polygon Buffer

Buffering points

A point is the basic unit of resolution in any GIS system. Buffering point data involves the creation of a circular polygon about the point of interest. The radius of this circular polygon is called the buffer distance. In this scheme the buffer distance or the radius of the circle could be fixed for all points in a layer or the user could specify it. If multiple points in the same layer are being buffered, then buffer distances of each point are either specified in an attribute table or a look up table. If one is buffering multiple points in the same layer, then the buffering algorithms check for overlaps in each point's buffer and remove the overlapping sections.

OBJECTID *	SHAPE *	buffer_field
1	Point	200 Meters
2	Point	1 Miles
3	Point	1 Kilometers
4	Point	500 Feet

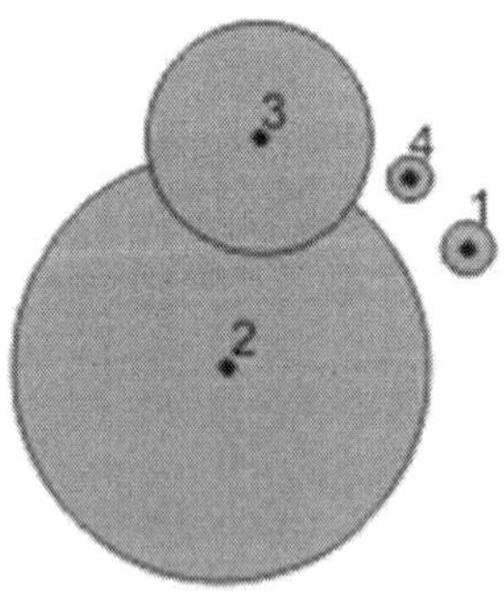

Figure 5.9 Point Buffer

Buffering Lines

Buffering lines is a little more complicated than buffering point data. This is mainly due to the fact that lines can be made up of multiple segments. Line segments are handled independently of each other.

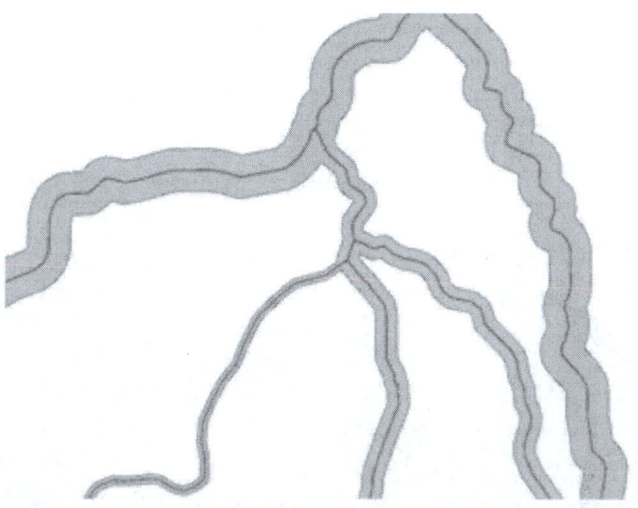

Figure 5.10 Line Buffer (See colour version on page 172)

Buffering Polygons

Buffering of polygonal surfaces uses most of the same concepts used for line buffering. The only significant change is that the polygon buffer is created on only one side of the line that defines the polygon. In polygon buffering two options are available, namely - an outside polygon which surrounds/contains the polygonal surface under consideration or an inside polygon that is contained inside the polygonal surface under consideration.

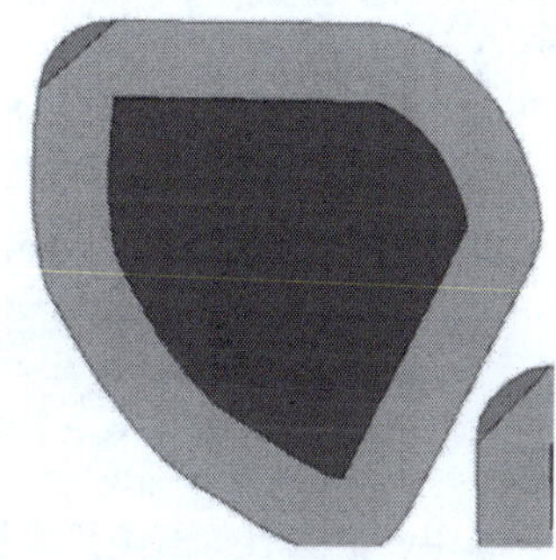

Figure 5.11 Area Buffer (See colour version on page 172)

Neighbourhood Operations

Neighbourhood operations evaluate the characteristics of an area surrounding a specific location. Virtually all GIS software provides some form of neighbourhood analysis. A range of different neighbourhood functions exist. The analysis of topographic features, e.g. the relief of the landscape, is normally categorized as being a neighbourhood operation. This involves a variety of *point interpolation* techniques

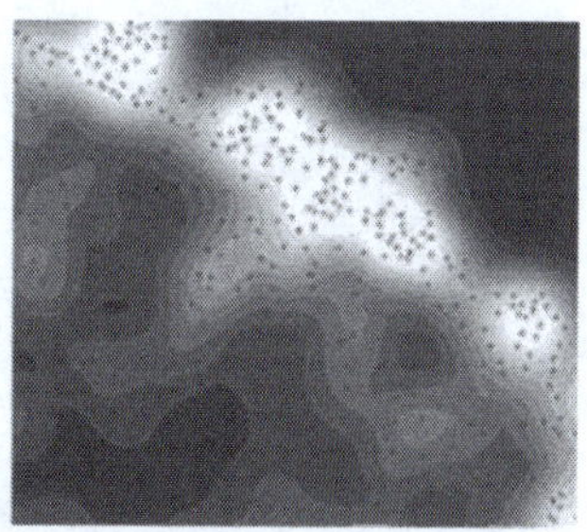

Figure 5.12 Neighbourhood operations (See colour version on page 172)

including slope and aspect calculations, contour generation, and *Thiessen polygons.* Interpolation is defined as the method of predicting unknown values using known values of neighbouring locations. Interpolation is utilized most often with point based elevation data.

This example illustrates a continuous surface that has been created by interpolating sample data points.

Elevation data usually takes the form of irregular or regular spaced points. Irregularly space points are stored in a Triangular Irregular Network (TIN). A TIN is a vector topological network of triangular facets generated by joining the irregular points with straight line segments. The TIN structure is utilized when irregular data is available, predominantly in vector based systems. TIN is a vector data model for 3-D data.

An alternative in storing elevation data is the regular point Digital Elevation Model (DEM). The term DEM usually refers to a grid of regularly space elevation points. These points are usually stored with a raster data model. Most GIS software offerings provide three dimensional analysis capabilities in a separate module of the software. Again, they vary considerably with respect to their functionality and the level of integration between the 3-D module and the other more typical analysis functions.

Without doubt the most common neighbourhood function is *buffering*. Buffering involves the ability to create distance buffers around selected features, be it points, lines, or areas. Buffers are created as polygons because they represent an area around a feature. Buffering is also referred to as ***corridor* or *zone generation*** with the raster data model. Usually, the results of a buffering process are utilized in a topological overlay with another data layer. For example, to determine the volume of timber within a selected distance of a cutline, the user would first buffer the cutline data layer. They would then overlay the resultant buffer data layer, a buffer polygon, with the forest cover data layer in a clipping fashion. This would result in a new data layer that only contained the forest cover within the buffer zone. Since all attributes are maintained in the topological overlay and buffering processes, a map or report could then be generated.

Buffering is typically used with point or linear features. The generation of buffers for selected features is frequently based on a distance from that feature, or on a specific attribute of that feature. For example, some features may have a greater zone of influence due to specific characteristics, e.g. a primary highway would generally have a greater influence than a gravel road. Accordingly, different size buffers can be generated for features within a data layer based on selected attribute values or feature types.

Connectivity Analysis

The distinguishing feature of connectivity operations is that they use functions that accumulate values over an area being traversed. Most often these include the analysis of surfaces and networks. Connectivity functions include *proximity analysis*, *network analysis*, spread functions, and three dimensional surface analysis such as *visibility* and *perspective viewing*. This category of analysis techniques is the least developed in commercial GIS software. Consequently, there is often a great difference in the functionality offered between GIS software offerings. Raster based systems often provide the more sophisticated surface analysis capabilities while vector based systems tend to focus on linear network analysis capabilities. However, this appears to be changing as GIS software becomes more sophisticated, and multi-disciplinary applications require a more comprehensive and integrated functionality. Some GIS offerings provide both vector and raster analysis capabilities. Only in these systems will one found a full range of connectivity analysis techniques.

Proximity analysis techniques are primarily concerned with the proximity of one feature to another. Usually *proximity* is defined as the ability to identify any feature that is near any other feature based on location, attribute value, or a specific distance. A simple example is identifying all the forest stands that are within 100 metres of a gravel road, but not necessarily adjacent to it. It is important to note that neighbourhood buffering is often categorized as being a proximity analysis capability. Depending on the particular GIS software package, the data model employed, and the operational architecture of the software it may be difficult to distinguish proximity analysis and buffering.

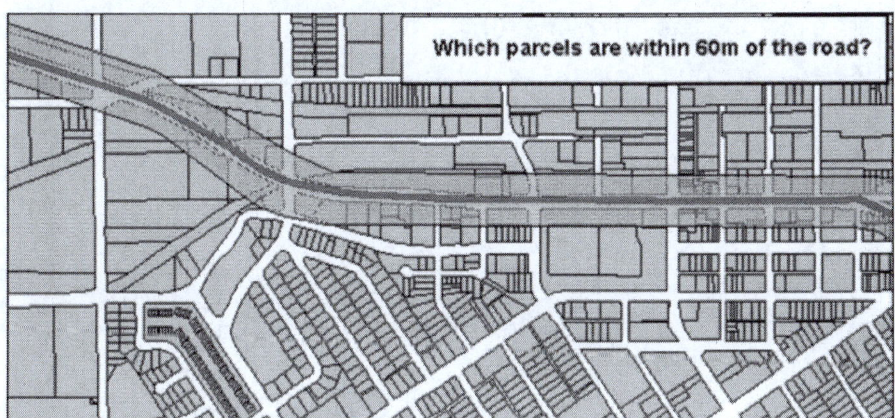

Figure 5.13 Proximity analysis

Proximity analysis is often used in urban based applications to consider areas of influence, and ownership queries. Proximity to roads and engineering infrastructure is typically important for development planning, tax calculations, and utility billing.

The identification of adjacency is another proximity analysis function. Adjacency is defined as the ability to identify any feature having certain attributes that exhibit adjacency with other selected features having certain attributes. A typical example is the ability to identify all forest stands of a specific type, e.g. specie, adjacent to a gravel road.

Network analysis is a widely used analysis technique. Network analysis techniques can be characterized by their use of feature networks. Feature networks are almost entirely comprised of linear features. Hydrographic hierarchies and transportation networks are prime examples. Two example network analysis techniques are the allocation of values to selected features within the network to determine capacity zones, and the determination of shortest path between connected points or nodes within the network based on attribute values. This is often referred to as route optimization. Attribute values may be as simple as minimal distance, or more complex involving a model using several attributes defining rate of flow, impedance, and cost.

Three dimensional analyses involve a range of different capabilities. The most utilized is the generation of perspective surfaces. Perspective surfaces are usually represented by a wire frame diagram reflecting profiles of the landscape, e.g. every 100 metres. These profiles viewed together, with the removal of hidden lines, provide a three dimensional view. As previously identified, most GIS software packages offer 3-D capabilities in a separate module. Several other functions are normally available.

These include the following functions

- User definable vertical exaggeration, viewing azimuth, and elevation angle;
- Identification of viewsheds, e.g. seen versus unseen areas;
- The draping of features, e.g. point, lines, and shaded polygons onto the perspective surface;
- Generation of shaded relief models simulating illumination;
- Generation of cross section profiles;
- Presentation of symbology on the 3-D surface; and
- Line of sight perspective views from user defined viewpoints.

Generation of DEM

What is DEM?

A digital elevation model (DEM) is a digital model or 3D representation of a terrain's surface – commonly for a planet (including Earth), moon, or asteroid – created from terrain elevation data.

Digital elevation model (DEM) is indispensable for many analyses such as topographic feature extraction, Runoff analysis, landslide susceptibility analysis and so on. Beforehand such analyses, accuracy of DEM must be discussed. The accuracy of DEM is usually represented by spatial resolution and height.

The DEM file also does not contain civil information such as roads or buildings. It is not a scanned image of the paper map (graphic). It is not a bitmap. The DEM does not contain elevation contours, only the specific elevation values at specific grid point locations. Delineation of terrain parameters, such as slope, drainage network, watershed boundaries etc., These parameters are often required in preparation of development and conservation plan for natural resources, infrastructure development, town planning, etc. DEM generated from stereo data has been validated using ground control points (GCPs). Evaluation of drainage pattern is carried out by morphometric analysis.

Digital Elevation Models (DEMs) are an important source of information in many geospatial applications where they are used in modelling, visualisation, correction of satellite imagery, etc. This is particularly the case at all stages of a mineral project from the exploration, development and mining stages. LiDAR surveys from aerial platforms are the most accurate however such surveys may be very expensive because of location charges or may not be available due to flying restrictions. In such circumstances, DEMs from Satellite Platforms are a viable option.

The following table includes the current satellites from which stereo imagery is available to produce DEMs as well as some coarse worldwide DEMs available from various sources and their accuracies.

Table 5.1 Dem data and accuracy

Source	Spatial resolution (m)	Accuracy (m)	Availability
SRTM	90	6-8	Land areas between latitudes 60 and -56
GDEM (ASTER)	30	Variable	Most land areas
SPOT HRS	20-30	8	From archive
WorldDEM	12	2m vert	Most land areas
ALOS PRISM	5	5	From archive
GeoDEM-2	2	1-2	New Capture
IKONOS	2	1-2	New Capture
WV-1, WV-2, GEOEYE-1	1	1	New Capture

The SRTM DEM is an interfrommetric DEM produced from radar data collected from a Shuttle mission in 2000. The GDEM is a DEM produced from stereo imagery collected from the ASTER sensor and processed semi-automatically by ERSDAC and JPL. The initial version of this DEM was released in mid 2009 and varies from very good to unuseable depending on the area. The HRS DEM was produced by SPOT IMAGE from an instrument that is flown on the SPOT5 satellite. The individual stereo imagery is not available for purchase.

Worldwide SRTM DEM

The PRISM sensor is an instrument on the Japanese ALOS satellite which started capturing imagery in May 2006. The sensor is unique in capturing back, nadir and forward looking images for a 35 km wide swath at 2.5 m resolution. A considerable archive of triplet imagery is available and can be processed to provide good quality DEMs.

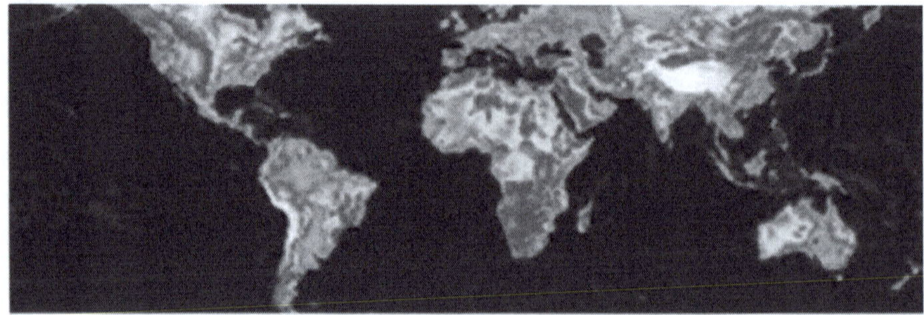

Figure 5.14 SRTM data (See colour version on page 173)

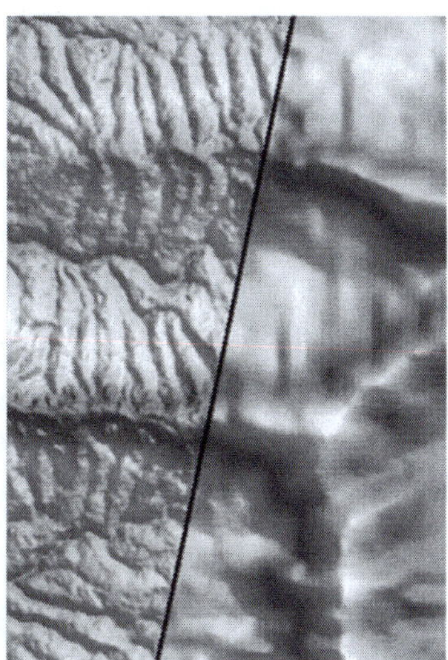

Figure 5.15 ALOS PRISM and SRTM data Comparison
(See colour version on page 173)

Comparison of ALOS PRISM DEM and SRTM

There are currently four Very High Resolution Satellites with ground resolution less than 1 m which can be programmed for in track stereo captures. These produce DEMs with a cell size of 1 to 2 m and accuracy in elevation of 1 to 2 m. Geoimage has found that the DEMs produced from the WorldView-2 satellite produce excellent quality DEMs.

Figure 5.16 DEM Model (See colour version on page 173)

Triangulated Irregular Network (TIN)

A **triangulated irregular network** (**TIN**) is a digital data structure used in a **geographic information system** (**GIS**) for the representation of a surface.

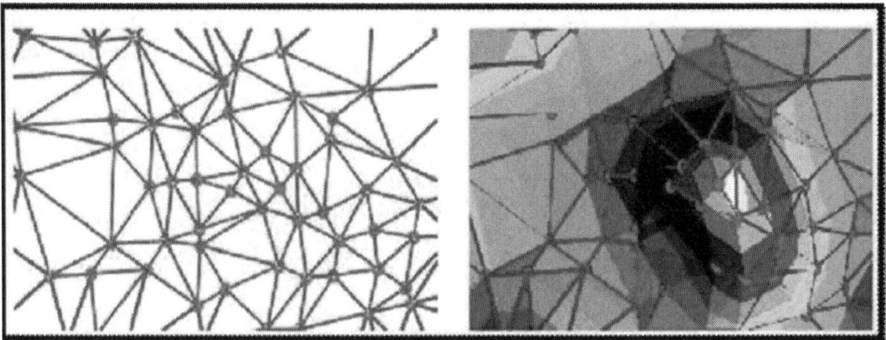

Figure 5.17 TIN Model (See colour version on page 173)

A **triangulated irregular network (TIN)** is a digital data structure used in a geographic information system (GIS) for the representation of a surface. A TIN is a vector-based representation of the physical land surface or sea bottom, made up of irregularly distributed nodes and lines with three-dimensional coordinates (x, y, and z) that are arranged in a network of non-overlapping triangles.

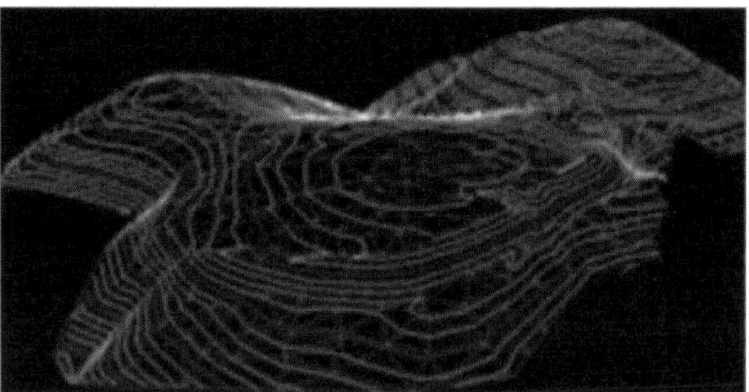

Figure 5.18 TIN overlaid with contour lines
(See colour version on page 174)

A TIN comprises a triangular network of vertices, known as mass points, with associated coordinates in three dimensions connected by edges to form a triangular tessellation. Three-dimensional visualizations are readily created by rendering of the triangular facets. In regions where there is little variation in surface height, the points may be widely spaced whereas in areas of more intense variation in height the point density is increased.

A TIN used to represent terrain is often called a Digital Terrain Model (DTM). An advantage of using a TIN over a rasterized digital elevation model (DEM) in mapping and analysis is that the points of a TIN are distributed variably based on an algorithm that determines which points are most necessary to an accurate representation of the terrain. Data input is therefore flexible and fewer points need to be stored than in a raster DEM, with regularly distributed points. A TIN may be less suited than a raster DEM for certain kinds of GIS applications, such as analysis of a surface's slope and aspect. A DTM can be formed from a DEM. A DEM can be interpolated from a TIN.

A TIN is typically based on a Delaunay triangulation, but its utility will be limited by the selection of input data points: well-chosen points will be located so as to capture significant changes in surface form, such as topographical summits, breaks of slope, ridges, valley floors, pits, and cols. Although usually associated with three-dimensional data (x, y, and z) and topography, TINs are also useful for the description and analysis of general horizontal (x and y) distributions and relationships.

Data Integration

Data integration involves combining data residing in different sources and providing users with a unified view of these data. This process becomes significant in a variety of situations, which include both commercial (when two similar companies need to merge their databases) and scientific combining research results from different bioinformatics repositories, for example domains. Data integration appears with increasing frequency as the volume and the need to share existing data explodes. It has become the focus of extensive theoretical work, and numerous open problems remain unsolved.

Definition of data integration

Data Integration is widely used in different areas of scientific circles and ordinary life as well. Data integration means combining Data coming from different sources and providing users with a unified view of these data. This process becomes significant in a variety of situations both commercial (when two similar companies need to merge their Databases) and scientific (combining research results from different repositories). Data integration appears with increasing frequency as the volume and the need to share existing data.

High degree of interconnection between two or more programs or datasets, in which they share a common schema, ontology, semantic approach, or method that allows information to be passed between them without being fully processed. We need to integrate many different data types for our analyses.

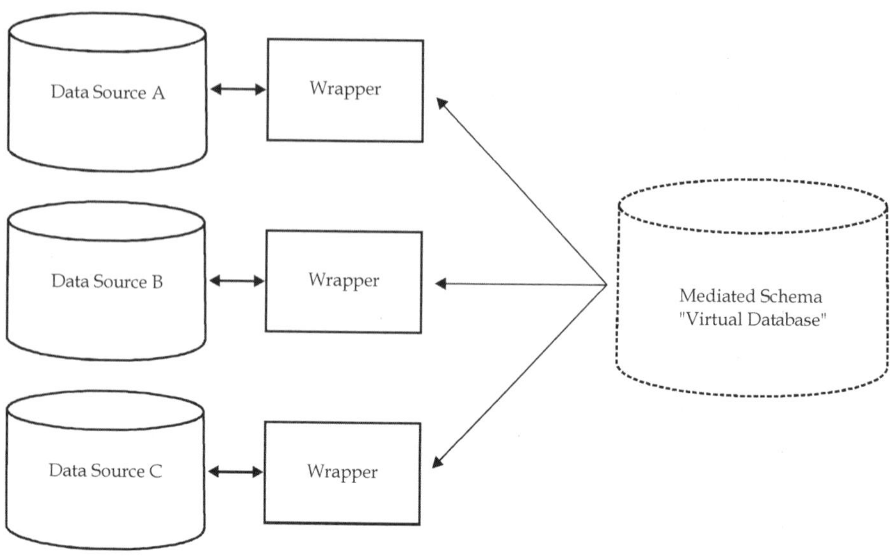

Figure 5.19 Schema of Data integration

A lot of theoretical work dealing with data integration, and numerous open problems remain unsolved. In 2010 some of the work in data integration research concerns the semantic integration problem. This problem focuses on resolving Semantic conflicts between heterogeneous data sources. For example if two companies merge their databases, certain concepts and definitions in their respective schemas have different meanings. A common strategy for the resolution of such problems involves the use of Ontologies (computer science) which explicitly define schema terms and thus help to resolve semantic conflicts. This approach represents ontology-based data integration.

The above mention methods and theories apply to data integration in GIS as well. In GIS the integration of data provides the ability to ask complex spatial questions that could not be answered otherwise.

Multiple use of existing data for full exploitation

If we have different data set the necessity of data conversion will be arise and then we should decide which conversion method to apply?

Decomposition of reality is consisting of different database. We can build databases with different entities, which mean that real word is modelled in way user can exploit data more effectively.

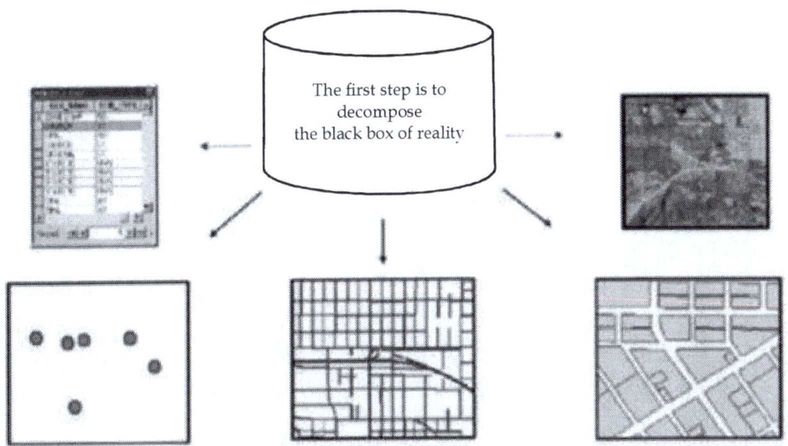

Figure 5.20 Store street data (*Source*: Guszlev A. : EDUGI course material) (See colour version on page 174)

We can decide what kind of GIS layers will represent our reality.

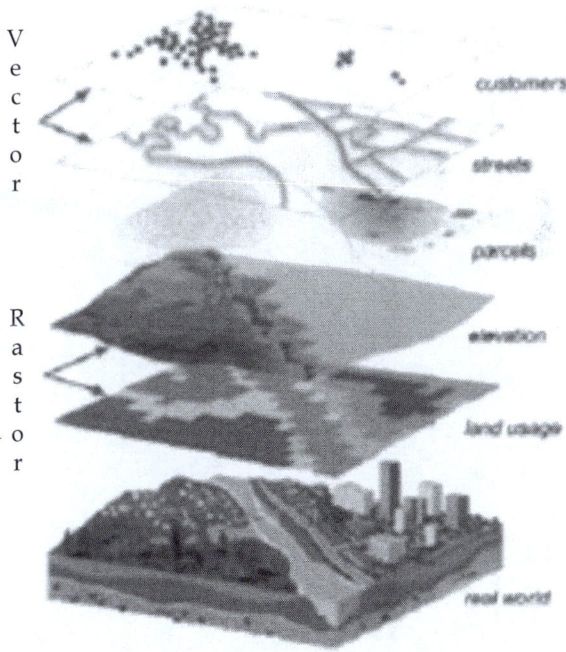

Figure 5.21 GIS Layer (See colour version on page 174)

The schema of reality with the layers of GIS (*Source*: Guszlev A. ; EDUGI course material).

Data conversion

If we would like to make some spatial analyses between different data (stored in different layers) we need to convert the different data model. It is our responsibility to decide which data model we would prefer. It depends on the aim of the project, the storing capacity, the available data.

Rasterization

Rasterization means the conversion of points, lines, and polygons into cell data. The term rastorization can in general be applied to any process by which Vector graphics information can be converted into a rastor format.

❑❑❑

6

Applications of Remote Sensing GIS and GPS

Applications of Remote Sensing

The era of satellite remote sensing was open in the 1960s when cameras and electronic sensors were mounted on spacecraft. Nowadays there is a big assortment of satellite systems actively recording information about the Earth. A wide variety of imagery is available from satellites. Both active and passive sensors, operating from the microwave to the ultraviolet regions of the electromagnetic spectrum collect a large amount of information about the earth's surface every day. Each of the systems vary in terms of their spatial, spectral, radiometic and temporal resolution. Those characteristics play an important role in defining which applications the sensor is best suited for. The main benefits of satellite remote sensing are the following:

- The data are available for large areas: for example: 35 000km^2 for LANDSAT scene, 3 600 km^2 for SPOT scene.
- They are available on a regular basis for all points on the globe (repetitive coverage): data may be acquired every 1-3 days (16 days in the case of LANDSAT, 1-3 days in the case of SPOT).
- They are objective: the sensor-transmission-reception system involves no human intervention.
- The data collected are related to the Earth surface features.
- To collect data the sensors use the wide spectrum of electromagnetic spectrum (EMS) and use several band (areas of the EMS) at once (LANDSAT TM 7, hyperspectral: from tens to hundreds of spectral bands).
- They are in digital form and geometrically corrected images can be used to provide a base to overlay other data or to be used as part of an analysis in a GIS environment.

The applications of remote sensing summarized in this unit are only representative, but not exhaustive. We focus on applications tied to the Earth surface. To show the remote sensing as data source and the benefits of remote sensing applications we emphasized some examples of touched fields.

Agricultural

Agriculture plays an important role in economies of countries. The production of food is important to everyone and producing food in a cost-effective manner is the goal of every farmer and an agricultural agency. The satellites have ability to image individual fields, regions and counties on a frequent revisit cycle. Customers can receive field-based information including **crop identification, crop area determination and crop condition monitoring** (health and viability). Satellite data are employed in precision agriculture to manage and monitor farming practices at different levels. The data can be used to farm optimization and spatially-enable management of technical operations. The images can help determine the location and extent of crop stress and then can be used to develop and implement a spot treatment plan that optimizes the use of agricultural chemicals. The major agricultural applications of remote sensing include the following:

Vegetation

- Crop type classification
- Crop condition assessment (crop monitoring, damage assessment)
- Crop yield estimation.

Soil

- Mapping of soil characteristics
- Mapping of soil type
- Soil erosion
- Soil moisture
- Mapping of soil management practices
- Compliance monitoring (farming practices).

Crop type classification

Remote sensing technology can be used to prepare maps of crop type and delineating their extent. Traditional methods of obtaining this information are census and ground surveying. The use of satellites is advantageous as it can generate a systematic and repetitive coverage of a large area and provide information about the health of the vegetation. The data of crop is needed for agricultural agencies to prepare an inventory of what was grown in certain areas and when. This information serves to predict grain crop yield, collecting crop production statistics, facilitating crop rotation records, mapping soil productivity, identification of factors influencing crop stress, assessment of crop damage and monitoring farming activity.

What kind of images can be used?

There are several types of remote sensing systems used in agriculture but the most common is a passive system that senses the electromagnetic energy reflected from plants. The spectral reflection of a vegetation depend on stage type, changes in the phenology (growth), and crop health, and thus can be measured and monitored by multi-spectral sensors. Many remote sensing sensors operate in the green, red, and near infrared regions of the EM spectrum, they measure both absorption and reflectance effects associated with vegetation. Multi-spectral variations facilitate fairly precise detection, identification and monitoring of vegetation. The observation of vegetation phenology requires multi-temporal images (data at frequent intervals throughout the growing season). Different sensors (multi-sensor) often provide complementary information, and when integrated together, can facilitate interpretation and classification of imagery. Examples include combining high resolution panchromatic imagery with coarse resolution multi-spectral imagery, or merging actively and passively sensed data (SAR imagery with multi-spectral imagery).

Figure 6.1 Some species (red color) of cultivated plant on satellite image (See colour version on page 175)

Crop Monitoring and Damage Assessment

Remote sensing has a number of attributes that lend themselves to monitoring the health of crops. The optical (VIR) sensing advantage is that it can see the infrared, where wavelengths are highly sensitive to crop vigor as well as crop stress and crop damage. Remote sensing imagery also gives the required spatial overview of the land. Remote sensing can aid in identifying crops affected by conditions that are too dry or wet, affected by insect, weed or fungal infestations or weather related damage (Fig. 6.2). Images can be obtained throughout the growing season to not only detect problems, but also to monitor the success of the treatment. Detecting damage and monitoring crop health requires high-resolution, multi-spectral imagery and multi-temporal imaging capabilities. One of the most critical factors in making imagery useful to farmers is a quick turnaround time from data acquisition to distribution of crop information.

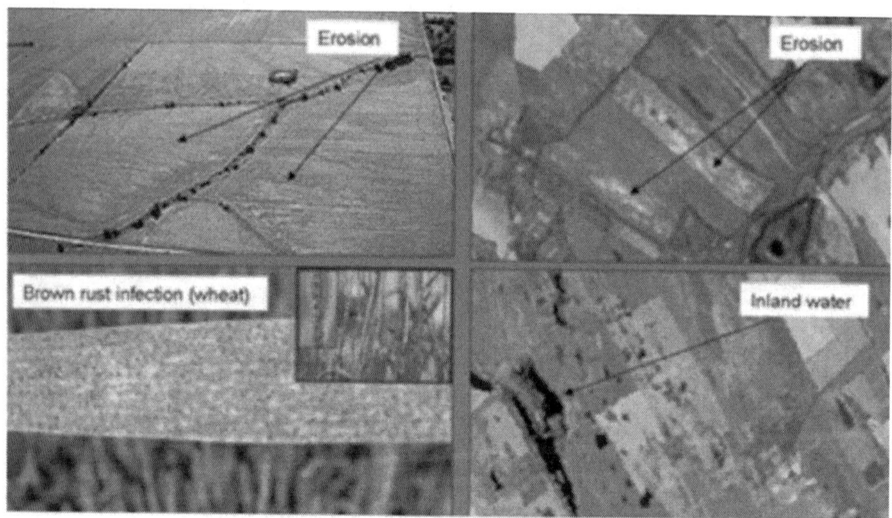

Figure 6.2 The problems inside the agricultural fields
(See colour version on page 175)

Soil mapping

The disturbance of soil by land use impacts on the quality of our environment. Salinity, soil acidification and erosion are some of the problems. Remote sensing is a good method for mapping and prediction of soil degradation. Soil layers that rise to the surface during erosion have different color, tone and structure than non eroded soils thus the eroded parts of soil can be easily identify on the images (Fig. 6.2). Using multi-temporal images we can study and map dynamical features - the expansion of erosion, soil moisture. Attempts to study land degradation processes and the necessity of degradation prediction have resulted in the creation of erosion models. The necessary information, parameters of the models; Universal Soil Loss Equation (USLE) to modeling can be often derived from satellite images. The vegetative cover is a major factor of soil erosion.

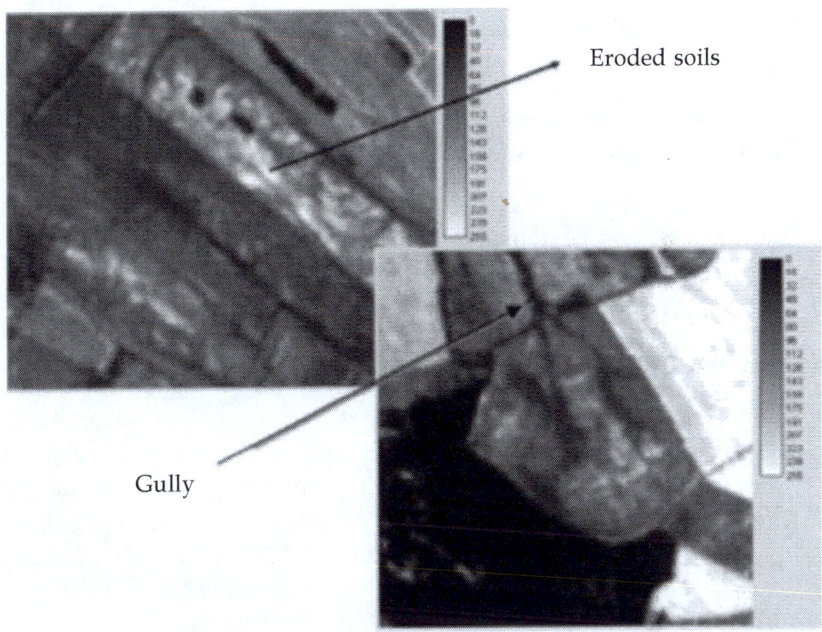

Figure 6.3 Some kind of erosion in satellite image

Forest mapping

One of the basic applications is forest cover typing and species identification. Forest cover typing can consist of reconnaissance mapping over a large area, while species inventories are highly detailed measurements of stand contents and characteristics (tree type, height, density). Using remote sensing data we can identify and delineate various forest types, that would be difficult and time consuming using traditional ground surveys. Data is available at various scales and resolutions to satisfy local or regional demands. Requirements for reconnaissance mapping depend on the scale of study. For mapping differences in forest cover (canopy texture, leaf density) are needed:

- Multi-spectral images, a very high resolution data is required to get detailed species identification.

- Multi-temporal images datasets contribute phenology information of seasonal changes of different species.
- Stereo photos help in the delineation and assessment of density, tree height and species.
- Hyperspectral imagery can be used to generate signatures of vegetation species and certain stresses (e.g. infestations) on trees. Hyperspectral data offers a unique view of the forest cover, available only through remote sensing technology.
- RADAR is more useful for applications in the humid tropics because its all weather imaging capability is valuable for monitoring forest.
- LIDAR data allows the 3-dimensional structure of the forest. The multiple return systems are capable of detecting the elevation of land and objects on it. The LIDAR data help estimate a tree height, a crown area and number of trees per unit area.

Clear cut Mapping and Deforestation

One of an important global problem is deforestation (Fig. 6.4). There are many implications of it: in industrialized parts of world, pollution (acid rain, soot and chemicals from factory smoke plumes) has damaged a large percentage of forested land, in tropical countries, valuable rainforest is being destroyed in an effort to clear potentially valuable agricultural and pasture land. The loss of forests increases soil erosion, river siltation and deposition, affecting the environment.

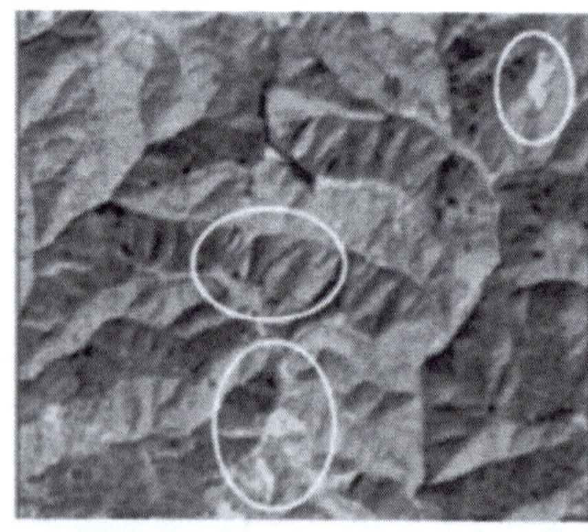

Figure 6.4 Forest cuts monitoring (LANDSAT TM 1992 and 2000).
Source: http://www.hso.hu/cgi-bin/page.php?page=84)
(See colour version on page 176)

Land cover

Land cover mapping is one of the most important and typical applications of remote sensing data. Land cover corresponds to the physical condition of the ground surface, for example, forest, grassland, concrete pavement etc., while land use reflects human activities such as the use of the land, for example, industrial zones, residential zones, agricultural fields etc. Initially the land cover classification system should be established, which is usually defined as levels and classes. The level and class should be designed in consideration of the purpose of use (national, regional or local), the spatial and spectral resolution of the remote sensing data, user's request and so on.

Land cover change detection is necessary for updating land cover maps and the management of natural resources. The change is usually detected by comparison between two multi-date images, or sometimes between an old map and an updated remote sensing image.

- **Seasonal change** : Agricultural lands and deciduous forests change seasonally
- **Annual change**: Land cover or land use changes, which are real changes, for example deforested areas or newly built towns.

Information on land cover and changing land cover patterns is directly useful for determining and implementing environment policy and can be used with other data to make complex assessments (e.g. mapping erosion risks).

CORINE Land Cover

CORINE Land Cover is the largest databases, providing information on the physical characteristics of the earth surface. Satellite images are used as the main source data to derive land cover information. The number of participating countries currently is 39. In the beginning of the project an information system on the state of the European environment was established, nomenclatures and methodologies were developed and agreed at European level. The aim of the European Union's CLC project was to map and get information on land cover at scale 1:100.000 for the whole Europe. In the beginning of 1990's the land cover of the contemporary European Union countries were mapped first time using satellite images, common classification criteria and quality standards. Classification was done by visual interpretation. In the 2000 (CLC2000) the majority of the participating countries updated the CLC1990 database using the same method of interpretation of the satellite images

(Fig. 6.5). In addition to CLC1990 and CLC2000 also the land cover changes (1990-2000) database was produced. CORINE Land Cover 2006 is the third European Land Cover inventory. Multi-temporal images (SPOT-4/5 and IRS-P6) were used to derive the minimum 5 ha land cover changes that occurred between 2000 and 2006. Some characteristics of the CORINE databases are as follows (Table 6.1):

Table 6.1 Some characteristics of the CORINE databases

Description	CORINE 1:100.000	CORINE 1:50.000
Satellite images	Landsat TM	Landsat ETM+ SPOT PAN
Date of images	1990-1992	1990-1993
Number of categories	44	87
Minimal extension of the area (ha)	25	4
Minimal width of the area (m)	100	50

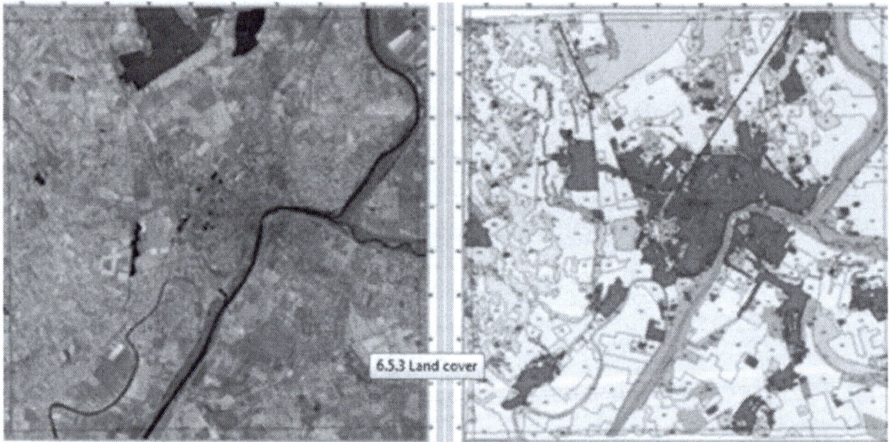

Figure 6.5 Spot satellite images as input data and the result of its interpretation. (*Source:* http://www.fomi.hu/corine/clc50.htm) (See colour version on page 176)

Remote sensing data has shown tremendous potential for applications in various field for example in land use mapping and detection, geologic mapping, water resource applications (pollution, lake-eutrophisation assessment), wetland mapping, urban and regional planning, environment inventory, natural disaster assessment or archaeological applications and other. In this unit we accented some examples of touched fields to show the remote sensing as data source and the benefits of remote sensing applications.

Indian Remote Sensing Application

Earth Observation

With a humble beginning in early 60s, Indian space program has matured as a symbol of the country's sophisticated technological capabilities and its growing regional and global prestige. Over the last four decades, Indian Space programme has made remarkable progress towards building the space infrastructure as the community resource to accelerate various developmental processes and harness the benefits of space applications for socio-economic development.

The Indian Space programme has the primary objective of developing space technology and application programmes to meet the developmental needs of the country. Towards meeting this objective, two major operational systems have been established - the Indian National Satellite (INSAT) for telecommunication, television broadcasting, and meteorological services and the Indian Remote Sensing Satellite (IRSS) for monitoring and management of natural resources and Disaster Management Support.

The Indian Remote sensing programme is driven by the user needs. In fact, the first remote sensing based pilot project was carried out to identify coconut root-wilt disease in Kerala way back in 1970. This pilot project led the development of Indian Remote Sensing (IRS) satellites. These IRS satellites have been the workhorse for several applications - encompassing the various sectors such as agriculture, land and water resources, forestry, environment, natural disasters, urban planning and infrastructure development, rural development, and forecasting of potential fishing zones.

A well knit network "Natural Resources Management System (NRMS)" involving Central & State Governments, private sectors, academia and Non-Governmental Organizations is in place for enabling the integration of Remote Sensing, contemporary technologies and conventional practices for management of natural resources.

Satellite Communication

The communication satellite series, which started with the APPLE satellite, grew into a very large constellation of satellites in the INSAT and GSAT series. These satellites revolutionized the technological and economic growth of the country. The INSAT satellite system is one of the largest domestic communication satellite systems providing regular services in the areas of telecommunications, business & personal communication, broadcasting, and weather forecasting & meteorological services. Today, newer initiatives have been taken to expand the INSAT applications to newer areas like Tele-education, Tele-medicine, Village Resource Centre (VRC), Disaster Management Support (DMS) etc., have enabled the space technology to reach the common man in India. The INSAT system has extended the outreach to less accessible areas like North- East, other far-flung areas and islands.

Indian Space Research Organisation (ISRO) has made remarkable progress towards building the space infrastructure - as the community resource to leapfrog the developmental processes. The launch of INSAT system has been the major catalyst in the rapid expansion of television coverage in India apart from growing applications like DTH, Satellite News Gathering, VSATs, Internet services etc. Use of INSAT for e-governance and developmental communication applications is also fast expanding.

Disaster Management Support

Disaster management support, in terms of space based critical infrastructure and services, is yet another community centric deliverable. One of the elements on which the space based Disaster Management Support (DMS) systems have been built is emergency communications systems. The DMS programme of ISRO/DOS, a convergence of space communications and remote sensing capabilities, is an effort to have technologically robust and a compatible system, which could strengthen India's resolves towards disaster management.

Satellite Navigation

Satellite Navigation service is an emerging satellite based system with commercial and strategic applications. ISRO is committed to provide the satellite based Navigation services to meet the emerging demands of the Civil Aviation requirements and to meet the user requirements of the positioning, navigation and timing based on the independent satellite navigation system. To meet the Civil Aviation requirements, ISRO is working jointly with Airport Authority of India (AAI) in establishing the GPS Aided Geo Augmented Navigation (GAGAN) system. To meet the user requirements of the positioning, navigation and timing services based on the indigenous system, ISRO is establishing a regional satellite navigation system called Indian Regional Navigation Satellite System (IRNSS).

Climate and Environment

ISRO has designed and developed indigenous systems for ground based observations of weather parameters. It includes (i) Automatic Weather Station (AWS) to providing hourly information on critical weather parameters such as pressure, temperature, humidity, rainfall, wind and radiation from remote and inaccessible areas; (ii) Agro Metrological (AGROMET) Towers to measure soil temperature, soil moisture, soil heat and net radiation, wind speed, wind direction, pressure and humidity; (iii) Flux Tower for multi-level micrometeorological observation as well as subsurface observations on soil temperature and moisture over the vegetative surfaces; (iv) Doppler Weather Radar (DWR) to monitor severe weather events such as cyclone and heavy rainfall; (v) GPS Sonde and Boundary Layer LIDAR (BLL) for observing vertical profiles of atmospheric parameters.

Applications of Geographic Information Systems (GIS)

Now-a-days, GIS technologies have been applied to diverse fields to assist experts and professionals in analyzing various types of geospatial data and dealing with complex situations. No matter in business, education, natural resources, tourism, or transportations, GIS plays an essential role to help people collect, analyze the related spatial data and display data in different formats. The following stories are the successful cases that benefit from applying Super GIS series products as their solutions in various fields.

Use of GIS has seen unprecedented growth in the last ten years. With the powerful technology getting cheaper and system memories expanding, meaning that we can handle much bigger sets of data, some say that GIS is in a golden age. It was once the preserve of the cartographer - few outside would have used it or needed it, yet recently GIS has become a core part of modern environmental science degrees. Geology, climatology, geography, statisticians, archaeology, oceanography, conservation based qualification and most other environmental sciences (1) now offer a module at undergraduate level. In the last five years, most universities have offered masters degrees specifically in GIS, or post-graduate certificates and diplomas with a prerequisite of environmental undergraduate degree. It is surprising to many that this technology has been around for over fifty years (2) because to people outside of the relevant fields, and endless possibilities. GIS is something that will make many jobs easier and faster and allow them to do more things in the same space of time with the click of a button. No longer are maps the exclusive preserve of the cartographer, now urban and rural developers, medical resource planners, conservation professionals, environment agency staff (to track and measure floodplains and the spread of protected species) archaeologists and utilities providers are just some of the environmentalist careers that can benefit from digital mapping.

1. GIS in Mapping

Mapping is a central function of Geographic Information System, which provides a visual interpretation of data. GIS store data in database and then represent it visually in a mapped format. People from different professions use map to communicate. It is not necessary to be a skilled cartographer to create maps. Google map, Bing map, Yahoo map are the best example for web based GIS mapping solution.

2. Telecom and Network services

GIS can be a great planning and decision making tool for telecom industries. GDi GISDATA enables wireless telecommunication organizations to incorporate geographic data in to the complex network design, planning, optimization, maintenance and activities. This technology allows telecom to enhance a variety of application like engineering application, customer relationship management and location based services.

3. Accident Analysis and Hot Spot Analysis

GIS can be used as a key tool to minimize accident hazard on roads, the existing road network has to be optimized and also the road safety measures have to be improved. This can be achieved by proper traffic management. By identifying the accident locations, remedial measures can be planned by the district administrations to minimize the accidents in different parts of the world. Rerouting design is also very convenient using GIS.

4. Urban Planning

GIS technology is used to analyze the urban growth and its direction of expansion, and to find suitable sites for further urban development. In order to identify the sites suitable for the urban growth, certain factors have to consider which is: land should have proper accessibility, land should be more or less flat, land should be vacant or having low usage value presently and it should have good supply of water.

5. Transportation Planning

GIS can be used in managing transportation and logistical problems. If transport department is planning for a new railway or a road route then this can be performed by adding environmental and topographical data into the GIS platform. This will easily output the best route for the transportation based on the criteria like flattest route, least damage to habitats and least disturbance from local people. GIS can also help in monitoring rail systems and road conditions.

6. Environmental Impact Analysis

EIA is an important policy initiative to conserve natural resources and environment. Many human activities produce potential adverse environmental effects which include the construction and operation of highways, rail roads, pipelines, airports, radioactive waste disposal and more. Environmental impact statements are usually required to contain specific information on the magnitude and characteristics of environmental impact. The EIA can be carried out efficiently by the help of GIS, by integrating various GIS layers, assessment of natural features can be performed.

7. Agricultural Applications

GIS can be used to create more effective and efficient farming techniques. It can also analyze soil data and to determine: what are the best crop to plant?, where they should go? how to maintain nutrition levels to best benefit crop to plant?. It is fully integrated and widely accepted for helping government agencies to manage programs that support farmers and protect the environment. This could increase food production in different parts of the world so the world food crisis could be avoided.

8. Disaster Management and Mitigation

Today a well-developed GIS systems are used to protect the environment. It has become an integrated, well developed and successful tool in disaster management and mitigation. GIS can help with risk management and analysis by displaying which areas are likely to be prone to natural or man-made disasters. When such disasters are identified, preventive measures can be developed.

9. Landslide Hazard Zonation using GIS

Landslide hazard zonation is the process of ranking different parts of an area according to the degrees of actual or potential hazard from landslides. The evaluation of landslide hazard is a complex task. It has become possible to efficiently collect, manipulate and integrate a variety of spatial data such as geological, structural, surface cover and slope characteristics of an area, which can be used for hazard zonation. The entire above said layer can well integrate using GIS and weighted analysis is also helpful to find Landslide prone area. By the help of GIS we can do risk assessment and can reduce the losses of life and property.

10. Determine Land use/Land cover changes

Land cover means the feature that is covering the barren surface. Land use means the area in the surface utilized for particular use. The role of GIS technology in land use and land cover applications is that we can determine land use/land cover changes in the different areas. Also it can detect and estimate the changes in the land use/ land cover pattern within time. It enables to find out sudden changes in land use and land cover either by natural forces or by other activities like deforestation.

11. Navigation (routing and scheduling)

Web-based navigation maps encourage safe navigation in waterway. Ferry paths and shipping routes are identified for the better routing. ArcGIS supports safe navigation system and provides accurate topographic and hydrographic data. Recently DNR's Coastal Resources Division began the task of locating, documenting, and cataloging these no historic wrecks with GIS. This division is providing public information that make citizens awareness of these vessel locations through web map. The web map will be regularly updated to keep the boating public informed of these coastal hazards to minimize risk of collision and injury.

12. Flood Damage Estimation

GIS helps to document the need for federal disaster relief funds, when appropriate and can be utilized by insurance agencies to assist in assessing monetary value of property loss. A local government need to map flooding risk areas for evaluate the flood potential level in the surrounding area. The damage can be well estimate and can be shown using digital maps.

13. Natural Resources Management

By the help of GIS technology the agricultural, water and forest resources can be well maintain and manage. Foresters can easily monitor forest condition. Agricultural land includes managing crop yield, monitoring crop rotation, and more. Water is one of the most essential constituents of the environment. GIS is used to analyze geographic distribution of water resources. They are interrelated, i.e. forest cover reduces the storm water runoff and tree canopy stores approximately 215,000 tons carbon. GIS is also used in afforestation.

14. GIS Solutions in Banking Sector

Today rapid development occurs in the banking sector. So it has become more market driven and market responsive. The success of this sector largely depends on the ability of a bank to provide customer and market driven services. GIS plays an important role providing planning, organizing and decision making.

15. Soil Mapping

Soil mapping provides resource information about an area. It helps in understanding soil suitability for various land use activities. It is essential for preventing environmental deterioration associated with misuse of land. GIS Helps to identify soil types in an area and to delineate soil boundaries. It is used for the identification and classification of soil. Soil map is widely used by the farmers in developed countries to retain soil nutrients and earn maximum yield.

16. GIS based Digital Taxation

In Local Governments, GIS is used to solve taxation problems. It is used to maximize the government income. For example, for engineering, building permits, city development and other municipal needs, GIS is used. Often the data collected and used by one agency or department can be used by another. Example Orhitec GIS Ltd., can supply you with a system to manage property tax on a geographic basis that can work interactively with the municipal tax collection department. Using GIS we can develop a digital taxation system.

17. Land Information System

GIS based land acquisition management system will provide complete information about the land. Land acquisition managements is being used for the past 3 or 4 years only. It would help in assessment, payments for private land with owner details, tracking of land allotments and possessions identification and timely resolution of land acquisition related issues.

18. Surveying

Surveying is the measurement of location of objects on the earth's surfaces. Land survey is measuring the distance and angles between different points on the earth surface. An increasing number of national and governments and regional organizations are using GNSS measurements. GNSS is used for topographic surveys where a centimeter level accuracy is provided. These data can be incorporated in the GIS system. GIS tools can be used to estimate area and also, digital maps can prepare.

19. Wetland Mapping

Wetlands contribute to a healthy environment and retain water during dry periods, thus keeping the water table high and relatively stable. During the flooding they act to reduce flood levels and to trap suspended solids and attached nutrients. GIS provide options for wetland mapping and design projects for wetland conservation quickly with the help of GIS. Integration with Remote Sensing data helps to complete wetland mapping on various scale. We can create a wetland digital data bank with spices information using GIS.

20. GIS Applications in Geology

Geologists use GIS in a various applications. The GIS is used to study geologic features, analyze soils and strata, assess seismic information, and or create three dimensional (3D) displays of geographic features. GIS can be also used to analyze rock information characteristics and identifying the best dam site location.

21. Detection of Coal Mine Fires

GIS technology is applied in the area of safe production of coal mine. Coal mine have developed an information management system, the administrators can monitor the safe production of coal mine and at the same time improve the abilities to make decisions. Fire happens frequently in coal mines. So it can assessed spontaneous combustion risk using GIS tools.(Kun Fang, GIS Network Analysis in Rescue of Coal Mine)

22. Assets Management and Maintenance

GIS helps organizations to gain efficiency even in the face of finite resources and the need to hold down the cost. Knowing the population at risk enables planners to determine where to allocate and locate resources more effectively. Operations and maintenance staff can deploy enterprise and mobile workforce. GIS build mobile applications that provide timely information in the field faster and more accurate work order processing.

23. GIS for Planning and Community Development

GIS helps us to better understand our world so we can meet global challenges. Today GIS technology is advancing rapidly, providing many new capabilities and innovations in planning. By applying known part of science and GIS to solve unknown part, that helps to enhance the quality of life and achieve a better future. Creating and applying GIS tools and knowledge allow us integrating geographic intelligence into how we think and behave.

24. GIS in Dairy Industry

Geographic Information System is used in a various application in the dairy industry, such as distribution of products, production rate, location of shops and their selling rate. These can be monitored by using GIS system. It can be also possible to understand the demand of milk and milk products in different region. GIS can prove to be effective tool for planning and decision making for any dairy industry. These advantages has added new vistas in the field of dairy farm and management.

25. Tourism Information System

GIS provides a valuable toolbox of techniques and technologies of wide applicability to the achievement of sustainable tourism development. This provide an ideal platform tools required to generate a better understanding, and can serve the needs of tourists. They will get all the information on click, measure distance, find hotels, restaurant and even navigate to their respective links. Information plays a vital role to tourists in planning their travel from one place to another, and success of tourism industry. This can bring many advantages for both tourist and tourism department.

26. Irrigation Water Management

Water availability for irrigation purposes for any area is vital for crop production in that region. It needs to be properly and efficiently managed for the proper utilization of water. To evaluate the irrigation performance, integrated use of satellite remote sensing and GIS assisted by ground information has been found to be efficient technique in spatial and time domain for identification of major crops and their conditions, and determination of their areal extent and yield. Irrigation requirements of crop were determined by considering the factors such as evapotranspiration, Net Irrigation Requirement, Field Irrigation Requirement, Gross Irrigation Requirement, and month total volume of water required, by organizing them in GIS environment. (A. M. Chandra, S.K. Ghosh, Remote Sensing and Geographical Information System).

27. Fire Equipment Response Distance Analysis

GIS can be used to evaluate how far (as measured as via the street network) each portion of the street network is from a firehouse. This can be useful in evaluating the best location for a new firehouse or in determining how well the fire services cover particular areas for insurance ratings.(Himachal Pradesh, Development Report)

28. Worldwide Earthquake Information System

One of the most frightening and destructive phenomena of nature is the occurrence of an earthquake. There is a need to have knowledge regarding the trends in earthquake occurrence worldwide. A GIS based user interface system for querying on earthquake catalogue will be of great help to the earthquake engineers and seismologists in understanding the behavior pattern of earthquake in spatial and temporal domain. (A.M. Chandra, S.K. Ghosh Remote Sensing and Geographical Information System).

29. Volcanic Hazard Identification

Volcanic hazard to human life and environment include hot avalanches, hot particles gas clouds, lava flows and flooding. Potential volcanic hazard zone can be recognized by the characteristic historical records of volcanic activities, it can incorporate with GIS. Thus an impact assessment study on volcanic hazards deals with economic loss and loss of lives and property in densely populated areas. The GIS based platforms enables us to find out the damage and rapid response against volcanic activities may helps to reduce the effect in terms of wealth and health of people.

30. Energy Use Tracking and Planning

GIS is a valuable tool that helps in the planning organizing and subsequent growth in the energy and utilities industries. The effective management of energy systems is a complex challenge. GIS has enormous potential for planning, design and maintenance of facility. Also it provide improved services and that too cost effectively.

31. GIS for Fisheries and Ocean Industries

GIS tools add value and the capability to ocean data. ArcGis is used to determine the spatial data for a fisheries assessment and management system. It is extensively used in the ocean industry area and we get accurate information regarding various commercial activities. To enhance minimizing cost for the fishing industry. Also it can determine the location of illegal fishing operations.

32. Monitor Rangeland Resources

GIS is a valuable tool used to monitor the changes of rangeland resource and for evaluating its impact on environment, livestock and wild life. Accurate observation and measurements are to be made to find out the changes in the rangeland conditions. GIS is also used to monitoring ecological and seasonal rangeland conditions.

33. Reservoir Site Selection

GIS is used to find a suitable site for the dam. GIS tries to find best location that respect to natural hazards like earthquake and volcanic eruption. For the finding of dam site selection the factors include economic factors, social considerations, engineering factors and environmental problems. This all information are layered in the GIS.

34. Forest Fire Hazard Zone Mapping

Forest is one of the important elements of the nature. It plays an important role in the local climate. Forest fires caused extensive damage to our communities and environmental resource base. GIS can effectively use for the forest fire hazard zone mapping and also for the loss estimation. GIS also help to capture real time monitoring of fire prone areas. This is achieved by the help of GNSS and satellite Remote Sensing.

35. Pest Control and Management

Pest control helps in the agricultural production. Increasing in the rate of pest and weeds can lead to decrease in the crop production. Therefore GIS plays an important role to map out infested areas. This leads in the development of weed and pest management plan.

36. Traffic Density Studies

GIS can effectively use for the management of traffic problems. Today's population along with the road traffic is increasing exponentially. The advantage of GIS makes it an attractive option to be used to face the emerging traffic problems. By creating an extensive database that has all the traffic information such as speed data, road geometry, traffic flow and other spatial data and processing this information will provide us the graphical bigger picture for the traffic management.

37. Deforestation

Now-a-days forest area is decreasing every year, due to different activities. GIS is used to indicate the degree of deforestation and vital causes for the deforestation process. GIS is used to monitor deforestation.

38. Space Utilization

GIS helps managers to organize and spatially visualize space and how it can best be used. Operational costs can be decreased by more efficiently using space including managing the moves of personal and assets as well as the storage materials. The 3D visualization in GIS platforms helps planers to create a feeling of experience like virtual walk inside the building and rooms before construction.

39. Desertification

Desertification is the land degradation due to climatic variations or human activities. GIS can provide the information of degraded land which can be managed by governmental agencies or by the communities themselves. GIS plays a vital role to reduce the desertification, the local governments are now widely depends on GIS for reducing desertification. With location based GIS analysis we can find where or which area is suitable for planting new vegetation and which area for the pipeline construction.

40. Disaster and Business Continuity Planning

Viewing building and locations assets along with emergency information such as weather patterns, and disaster zones, can provide organizations the required information to make better decision. GIS provide holistic understanding of facility status and performance, and brings together department, business systems, and data source for a comprehensive view into and throughout the organization. (Faisal I. Al-Shukri, GIS Utilization in Facility Management).

41. GIS for Business

GIS is also used for managing business information based on its location. GIS can keep a track of where the customers are located, site business, target marketing campaigns, and optimize sales territories and model retail spending patterns. Such an added advantage is provided by the GIS to enhance in making companies more competitive and successful. (Sita Mishra, GIS in Indian Retail Industry-A Strategic Tool)

42. Utilities

The GIS is used for different type of utilities like electricity, telecom and cooking gas on a daily basis and utilities to help them in mapping, in inventory systems, track maintenance, monitor regulatory compliance or model distribution analysis, transformer analysis and load analysis.

43. Lease Property and Management

Revenue can be increased, operations and maintenance cost can be reduced when GIS is used to help manage space. Real estate and property managers can see and make queries about space including its availability, size and special constraints for the most cost effective use.

44. Development of Public Infrastructure Facilities

GIS has many uses and advantages in the field of facility management. GIS can be used by facility managers for space management, visualization and planning, emergency and disaster planning and response. It can be used throughout the life cycle of a facility from deciding where to build to space planning. Also it provides facilitate better planning and analysis.(Gulzara Mamazhakypova)

45. GIS for Drainage Problems in Tea Plantation Areas

Drainage problem in tea plantation differ widely because of its varied nature of physical conditions. Tea crop requires moisture at adequate levels all times of its growth. Any variation either excess or lack has a direct impact on the tea yield. This become greatly influenced the productivity of tea. Required some hydraulic design to solve this problem such as design of drains, checking the adequacy of the river and classification of water logged areas etc. GIS helps to reduce the water-logging by establishing well developed plans.

46. Collection of Information about Geographic Features

GIS is not simply a computer system used for making maps. A map is simply the most common way of reporting information from a GIS database. So these systems are not only for creating maps but also most importantly the collection of information about the geographic features such as building, roads, pipes, streams, ponds and many more that are located in your community.

47. GIS for Public Health

GIS provides the cost effective tool for evaluating interventions and policies potentially affecting health outcomes. GIS analysis, environmental health data is also helpful in explaining disease patterns of relationships with social, institutional, technological and natural environment. It can be understand the complex spatial temporal relationship between environmental pollution and disease, and identifying exposures to environmental hazards. GIS can significantly add value to environmental and public health data (The application of GIS in environmental health sciences: opportunities and limitations).

48. Location Identification

This technique is used to find a location for a new retail outlet. It helps to find out what exists at a particular location. A location can be described in many ways, using, for instance, name of place, post code, or geographic reference such as longitude or latitude or X/Y.

49. Knowledge Based System for Defense Purpose

Regular analysis of terrain is essential for today's fast paced battlefield. Conventional method of studying paper topographical maps is being replaced by use of maps in digital form to get terrain information. It is increasingly being used to derive terrain information from digital images. Which help to the selection of suitable sites for various military uses more accurate and faster. The uses of GIS provide information regarding the terrain features which can be useful for planning today's war strategies.

50. Pipeline Route Selection

Pipeline route planning and selection is usually a complex task. GIS technology is faster, better and more efficient in this complex task. Accurate pipeline route selection brings about risk and cost reduction as well as better decision making process. GIS least cost path analysis have been effectively used to determine suitable oil and gas pipeline routes. An optimal route will minimize reduce economic loss and negative socio-environmental impacts.

51. Producing Mailing Labels for Abutter Notification

Zoning board of appeals hearings or proposed action by a town or city require notifying abutting property owners. A GIS application for producing abutter mailing labels enables you to identify abutting property owners are in different ways. Once the properties are identified this kind of GIS applications can produce mailing labels and be integrated with a word processing "mail merge "capabilities. (http://www.mass.gov/)

52. Site Suitability for Waste Treatment Plant

There is an increasing amount of waste due to the over population growth. This has negative impact on the environment. With the help of GIS we can integrate various aspect layers in GIS and can identify which place is suitable for waste treatment plant. This process will reduce the time and it is cost effective. Also it enhances the accuracy. It provides a GIS analyst to identify a list of suitable dumping sites for further investigations. It also provides a digital bank for future monitoring program of the site.

53. Geologic Mapping

GIS is an effective tool in geological mapping. It becomes easy for surveyors to create 3D maps of any area with precise and desired scaling. The results provide accurate measurements, which helps in several field where geological map is required. This is cost effective and offers more accurate data, there by easing the scaling process when studying geologic mapping.

54. Environment

The GIS is used every day to help protect the environment. The environmental professional uses GIS to produce maps, inventory species, measure environmental impact, or trace pollutants. The environmental applications for GIS are almost endless. It can be used to monitor the environment and analyze changes.

55. Infrastructure Development

Advancement and availability of technology has set a new mark for professionals in the infrastructure development area. Now more and more professionals are seeking help of these technologically smart and improved information systems like GIS for infrastructure development. Each and every phase of infrastructure life cycle is greatly affected and enhanced by the enrollment of GIS.

56. Coastal Development and Management

The coastal zone represents varied and highly productive ecosystem such as mangrove, coral reefs, seagrasses and sand dunes. GIS could be generating data required for macro and micro level planning of coastal zone management. GIS could be used in creating baseline inventory of mapping and monitoring coastal resources, selecting sites for brackish water aquaculture, studying coastal land forms.

57. Crime Analysis

GIS is a necessary tool for crime mapping in law enforcement agencies worldwide. Crime mapping is a key component of crime analysis. Satellite images can display important information about criminal activities. The efficiency and the speed of the GIS analysis will increase the capabilities of crime fighting.

58. River Crossing Site Selection for Bridges

The important geotechnical consideration is the stability of slope leading down to and up from the water crossing. It is advisable to collect historical data on erosion and sedimentation. On the basis of these information asses the amount of river channel contraction, degree of curvature of river bend, nature of bed and bank materials including the flood flow and the flow depth, all these can be done in GIS within estimated time and accurately. This information has been often used for river crossing site selection for bridges.

59. Land Use Changes Associated with Open Cast Strip Mining

Mining is the back bone of the developing economy of any country. Mapping, monitoring and controlling the impact caused by the mining activities is necessary so as to understand the character and magnitude of these hazardous events in an area. The data required to understand the impact of mining from the environment is coming from different discipline, which need integration in order to arrive hazard map zonation.

60. Economic Development

GIS technology is a valuable tool used for the economic development. It helps in site selection, suitability analysis, and for finding the right sites to locate new business and grow existing ones. Within economic development, GIS is used to support the emerging trend of economic gardening, a new way to foster local and regional economic growth by existing small business in the community (By Ahmed Abukhater, GIS for Planning and Community Development: Solving Global Challenges).

61. School Student Walking Distance Analysis

If your community buses students to school, but only if they lived beyond a certain distance from their school, this can be used to determine what addresses are eligible for busing.

62. Locating Underground Pipes and Cables

Pipe line and cable location is essential for leak detection. It can be used to understand your water network, conducting repairs and adjustments, locating leaks known distance for correlating etc. Pipelines are continually monitored, check for leak detection and avoid the problem of geohazards.

63. Coastal Vegetation Mapping and Conservation

Coastal vegetation like Mangroves are the protectors of coast from natural hazards like tsunami, so that the conservation of these vegetation are highly important. GIS enable us to map which are having higher density of vegetation and which area need more vegetated? Integration of these details to coastal zone mapping helps to identify the area prone to coastal erosion and we can plant more vegetation to reduce coastal erosion.

64. Regional Planning

Every day, planners use Geographic Information System (GIS) technology to research, develop, implement, and monitor the progress of their plans. GIS provides planners, surveyors, and engineers with the tools they need to design and map their neighborhoods and cities. Planners have the technical expertise, political savvy, and fiscal understanding to transform a vision of tomorrow into a strategic action plan for today, and they use GIS to facilitate the decision-making process. (ESRI, GIS Solutions for Urban and Regional Planning)

65. GIS for Land Administration

In a number of countries, the separate functions of land administration are being drawn together through the creation of digital cadastral databases, with these database they can reuse land for suitable needs, digital taxation and even utilities are also easily handle using these database.

66. Snow Cover Mapping and Runoff Prediction

Systematic, periodical and precise snow cover mapping supported by GIS technology, and the organization of the results in a snow cover information system forms the basis for a wide range of applications. On the practical side, these applications are related to the monitoring of seasonal and yearly alterations of the snow cover under the presently existing climatic conditions, to simulate and forecast runoff, to map the regional distribution of the water equivalent, and to document the recession process of the snow cover during the melting period in its relation to geological features.

67. GIS for Wildlife Management

Manmade destruction such as habitat loss, pollution, invasive species introduction, and climate change, are all threats to wildlife health and biodiversity. GIS technology is an effective tool for managing, analyzing, and visualizing wildlife data to target areas where international management practices are needed and to monitor their effectiveness. GIS helps wildlife management professionals examine and envision.

Applications of Global Positioning System (GPS)

The applications of the Global Positioning System fall into five categories: location, navigation, timing, mapping, and tracking. Each category contains uses for the military, industry, transportation, recreation and science.

GPS is an essential element of the global information infrastructure. The free, open, and dependable nature of GPS has led to the development of hundreds of applications affecting every aspect of modern life. GPS technology is now in everything from cell phones and wristwatches to bulldozers, shipping containers, and ATM's.

Location

This category is for position determination and is the most obvious use of the Global Positioning System. GPS is the first system that can give accurate and precise measurements anytime, anywhere and under any weather conditions. Some examples of applications within this category are:

1. Measuring the movement of volcanoes and glaciers.
2. Measuring the growth of mountains.
3. Measuring the location of icebergs - this is very valuable to ship Captains helping them to avoid possible disasters.
4. Storing the location of where you were - most GPS receivers on the market will allow you to record a certain location. This allows you to find it again with minimal effort and would prove useful in a hard to navigate place such as a dense forest.

Navigation

Navigation is the process of getting from one location to another. And for this the Global Positioning System was designed for: The GPS system allows us to navigate on water, air, or land. It allows planes to land in the middle of mountains and helps medical evacuation helicopters save precious time by taking the best route.

Timing

GPS brings precise timing to all us. Each satellite is equipped with an extremely precise atomic clock. This is why we can all synchronize our watches so well and make sure international events are actually happening at the same time.

Mapping

This is used for creating maps by recording a series of locations. The best example is surveying where the DGPS technique is applied but with a twist. Instead of making error corrections in real time, both the stationary and moving receivers calculate their positions using the satellite signals. When the roving receiver is through making measurements, it then takes them back to the ground station which has already calculated the errors for each moment in time. At this time, the accurate measurements are obtained.

Tracking

The applications in this category are ways of monitoring people and things such as packages. This has been used along with wireless communications to keep track of some criminals. The suspect agrees to keep a GPS receiver and transmitting device with him at all times. If he goes where he's not allowed to, the authorities will be notified. This can also be used to track animals.

Many applications use a combination of the above categories. The following is a list of some additional areas where the global positioning service is being applied:

1. Precision farming
2. Open-Pit Mining
3. Oil exploration
4. Airport and harbour approaches
5. Animal migration and population studies
6. Vehicle tracking
7. Construction - tunnels, golf courses, roads, etc.
8. Emergency services - the closest ambulance or fire truck is sent to an emergency, thereby saving time
9. Atmospheric studies - ozone layer, air quality, etc.
10. Archaeological explorations
11. Recreational activities - camping, boating, etc.
12. Astronomical telescope pointing
13. Networking - speed is increased, the Internet is highly synchronized because of GPS
14. Aiding the blind.

Open-pit Mining

This is an area where very high precision is needed, regular DGPS techniques are not accurate enough. DGPS improves the accuracy from hundreds of meters down to less than 10 meters, but a technique called Real Time Kinematic (GPS) can increase the accuracy to within several centimeters. This RTK technique involves not only receiving signals from satellites but also receives correction information from nearby reference points.

RTK GPS is used in many other areas other than mining, but why does open-pit mining require such high accuracy? It can be used for measuring how steep a road should be, saving construction time. It can also be used for high accuracy with drills and shovels. The main reason of the high accuracy is needed for ore control to make sure the different levels of ore containing dirt go to the right places. Not all mines have the equipment to use RTK GPS yet, but it may become necessary as the minerals become less abundant and better techniques must be applied.

The technology of the Global Positioning System is allowing for huge changes in society. The applications using GPS are constantly growing. The cost of the receivers is dropping while at the same time the accuracy of the system is improving. This affects everyone with things such as faster Internet speed and safer plane landings.

Even though the system was originally developed for military purposes, civil sales now exceed military sales.

GPS Update

On May 1, 2000, President Bill Clinton announced that the government wills no longer scramble signals from the GPS satellites. This means that civilians will be able to enjoy the high accuracy that the military has had for years. The DGPS techniques and the equipment needed to use them will no longer be necessary to get the same effects on accuracy. The affects on society will be: more reliable and accurate measurements less costly to corporations - since the error removing equipment will no longer be needed more affordable and accessible to consumers greater incentive for the development of new uses. The accuracy should increase ten-fold, from 100 meter to 10 meter accuracy. The 8 billion U.S. dollar a year GPS industry should also see a much larger than expected increase in sales.

Indian Regional Navigation Satellite System (IRNSS)

IRNSS is an independent regional navigation satellite system being developed by India. It is designed to provide accurate position information service to users in India as well as the region extending up to 1500 km from its boundary, which is its primary service area. An Extended Service Area lies between primary service area and area enclosed by the rectangle from Latitude 30° South to 50° North, Longitude 30° East to 130° West.

IRNSS will provide two types of services, namely, Standard Positioning Service (SPS) which is provided to all the users and Restricted Service (RS), which is an encrypted service provided only to the authorised users. The IRNSS System is expected to provide a position accuracy of better than 20 m in the primary service area.

Some applications of IRNSS are:

- Terrestrial, Aerial and Marine Navigation
- Disaster management
- Vehicle tracking and fleet management
- Integration with mobile phones
- Precise timing
- Mapping and Geodetic data capture
- Terrestrial navigation aid for hikers and travellers
- Visual and voice navigation for drivers.

The IRNSS Signal-in-Space Interface Control Document (ICD) for Standard Positioing Service (SPS) is released to the public to provide the essential information on the IRNSS signal-in-space, to facilitate research & development and aid the commercial use of the IRNSS signals for navigation-based applications. Click here to Register and Download the Document.

The space segment consists of the IRNSS constellation of seven satellites, NavIC. Three satellites are located in suitable orbital slots in the geostationary orbit and the remaining four are located in geosynchronous orbits with the required inclination and equatorial crossings in two different planes. All the satellites of the constellation are configured identically. The satellites are configured with I-1K Bus to be compatible for launch on-board PSLV.

IRNSS-1A

IRNSS-1A was the first navigational satellite in the Indian Regional Navigation Satellite System series of satellites to be placed in geosynchronous orbit. It was built at ISRO Satellite Centre, Bangalore, costing Rs.125 crore (US$19 million). It has a lift-off mass of 1380 kg, and carries a navigation payload and a C-band ranging transponder, which operates in L5 band (1176.45 MHz) and S band (2492.028 MHz). An optimised I-1K bus structure with a power handling capability of around 1600 watts is used and is designed for a ten-year mission. The satellite was launched on-board PSLV-C22 on 1 July 2013 from Satish Dhawan Space Centre at Sriharikota.

IRNSS-1B

IRNSS-1B is the second out of seven in the Indian Regional Navigation Satellite System. It was very precisely and successfully placed in its orbit through PSLV-C24 rocket on 4 April 2014.

IRNSS-1C

IRNSS-1C is the third out of seven in the Indian Regional Navigation Satellite System series of satellites. The satellite was successfully launched using India's PSLV-C26 from Satish Dhawan Space Centre at Sriharikota on 16 October 2014 at 1:32 A.M.

IRNSS-1D

IRNSS-1D is the fourth out of seven in the Indian Regional Navigation Satellite System series of satellites system. It was successfully launched using India's PSLV-C27 on 28 March 2015 at 5:19 P.M.

IRNSS-1E

IRNSS-1E is the fifth out of seven in the Indian Regional Navigation Satellite System series of satellites system. It was successfully launched on 20 January 2016 using India's PSLV-C31 at 9:31 A.M.

IRNSS-1F

IRNSS-1F is the sixth out of seven in the Indian Regional Navigation Satellite System series of satellites system. It was successfully launched on 10 March 2016 using India's PSLV-C32 at 4:01 P.M.

IRNSS-1G

IRNSS-1G is the seventh and final of the IRNSS series of satellites system. It was successfully launched on 28 April 2016 at 12:51 P.M. IST using India's PSLV-C33. The successful launch of this satellite concludes the setting up of India's first indigenous Regional Navigation Satellite System.

□□□

Glossaries for Geospatial Technologies

REMOTE SENSING

Absorption band - Wavelength within which electromagnetic radiation is absorbed by the atmosphere.

Active remote Sensing - RS methods that provide their own source of EMR to illuminate the terrain (example: radar).

Atmospheric window - Wavelength within which the atmosphere readily transmits EMR.

Classification - Process of assigning individual pixels of an image to categories based on spectral reflectance characteristics.

Color infrared photography or imagery - Visible and near infrared wavelengths the "chlorophyll detector", useful for vegetation studies.

Digital data - Recorded as a number for each picture element (pixel).

Displayed as a black and white photo or image.

Electromagnetic spectrum - The ordered array of electromagnetic radiation extending from short cosmic waves to long radio waves.

Focal length - In cameras, the distance from the optical center of the lens to the plane at which the image of a very distant object is brought into focus.

Multispectral - Refers to remote sensing in two or more bands or wavelengths.

Multispectral is high spectral resolution (multiple bands).

Nadir - Point on the ground directly in line with the RS system and the center of the earth (directly below the sensor).

Near infrared bands generally 0.7 to 1.1 μm, used for vegetation discrimination, chlorophyll detection, used in combination with visible bands to produce a CIR image.

Panchromatic - One band covering all the visible bands.

Panchromatic is low spectral resolution (one band).

Parallax – Displacement of the position of a target in an image caused by a shift in the observation position...offset and layover.

Passive remote sensing instruments which measure reflectance from the sun (visible through IR).

Pixel - Picture element, a grid cell in digital imagery.

Radiant energy peak – Wavelength at which the maximum EM energy is radiated at a particular temperature. 0.49 μm and 9.7 μm.

Radiometric resolution - Number of gray levels, e.g. 0-64, 0-255, 0-2048 etc.

Remote sensing - The measurement or acquisition or information of an object by a recording device that is not in physical contact with the object under study, that is, sensing from a distance.

Resolution – The ability to separate closely spaced objects on an image or photo.

Scale - The ratio of a distance on an image or map to its corresponding distance on the ground. Scale = map distance/ground distance.

Scattering – Multiple reflections (in all directions) of EM waves by particles or surfaces.

Signature – Set of characteristics (often spectral) by which a material or an object may be identified on an image or photo.

Spatial resolution - Size of a pixel, determines amount of detail that visible on a image.

Spectral band - An interval in the electromagnetic spectrum (aka wavelength) recorded by a remote sensing instrument.

Spectral reflectance – The reflectance measured within a specific wavelength.

Spectral resolution - Width and number of spectral bands to which a sensor is sensitive.

Temporal resolution - How often data is collected over a particular location.

Visible wavelengths - The radiation range in which the human eye is sensitive, approximately 0.4 to 0.7 μm.

Wavelength – Distance between successive wave crests.

GIS

Attribute - A characteristic of a geographic feature, typically stored in tabular format and linked to the feature in a relational database. The attributes of a well-represented point might include an identification number, address, and type.

Base Layer - A primary layer for spatial reference, upon which other layers are built. Examples of a base layer typically used are either the parcels, or street centerlines.

Buffer - A zone of a specified distance around a feature.

Computer Aided Design (CAD) - An automated system for the design, drafting and display of graphically oriented information.

Coordinate - An x,y location in a Cartesian coordinate system or an x,y,z coordinate in a three dimensional system. Coordinates represent locations on the Earth's surface relative to other locations.

Database - A logical collection of interrelated information, managed and stored as a unit. A GIS database includes data about the spatial location and shape of geographic features recorded as points, lines, and polygons as well as their attributes.

Digital Elevation Model (DEM) - Terrain elevation data provided in digital form.

Digitize - To encode map features as x,y coordinates in digital form. Lines are traced to define their shapes. This can be accomplished either manually or by use of a scanner.

Geocode - The process of identifying a location by one or more attributes from a base layer.

Geographic Information System (GIS) - An organized collection of computer hardware, software, geographic data, and personnel designed to efficiently capture, store, update, manipulate, analyze, and display all forms of geographically referenced information.

Global Positioning System (GPS) - A satellite based device that records x,y,z coordinates and other data. Ground locations are calculated by signals from satellites orbiting the Earth. GPS devices can be taken into the field to record data while walking, driving, or flying.

Layer - A logical set of thematic data described and stored in a map library. Layers act as digital transparencies that can be laid atop one another for viewing or spatial analysis.

Line – Lines represent geographic features too narrow to be displayed as an area at a given scale, such as contours, street centerlines, or streams.

Metadata – Information about a data set. It may include the source of the data; its creation date and format; its projection, scale, resolution, and accuracy; and its reliability with regard to some standard.

Ortho Imagery - Aerial photographs that have been rectified to produce an accurate image of the Earth by removing tilt and relief displacements, which occurred when the photo was taken.

Point – A single x,y coordinate that represents a geographic feature too small to be displayed as a line or area at that scale.

Polygon - A multisided figure that represents area on a map. Polygons have attributes that describe the geographic feature they represent.

Scale – The ratio or relationship between a distance or area on a map and the corresponding distance or area on the ground.

Spatial Analysis - The process of modeling, examining, and interpreting model results. Spatial analysis is useful for evaluating suitability and capability, for estimating and predicting, and for interpreting and understanding.

Structured Query Language (SQL) - A syntax for defining and manipulating data from a relational database. Developed by IBM in the 1970s, it has become an industry standard for query languages in most relational database management systems.

Theme – An ArcView theme stores map features as primary features (such as arcs, nodes, polygons, and points) and secondary features such as tics, map extent, links, and annotation. A theme usually represents a single geographic layer, such as soils, roads, or land use.

GPS

C/A code -The standard (Course/Acquisition) GPS code. A sequence of 1023 pseudo-random, binary, biphase modulations on the GPS carrier at a chip rate of 1.023 MHz. Also known as the "civilian code."

Control segment - A world-wide network of GPS monitor and control stations that ensure the accuracy of satellite positions and their clocks.

Differential positioning - Accurate measurement of the relative positions of two receivers tracking the same GPS signals.

DGPS - Differential GPS.

Ephemeris - The predictions of current satellite position that are transmitted to the user in the data message. A table given for successive days the positions of heavenly bodies.

GLONASS - GLObal NAvigation Satellite System - Russian.

GPS - Global Positioning System.

Latitude - The location on the Earth measuring how far north or south of the equator one is.

Longitude - The location on the Earth measured east or west.

LORAN - LOng RAnge Navigation.

Nautical mile - Length measurement used in navigation and is 1/60 of 1 degree of the equator. One nautical mile is 6,080.2 feet whereas one mile is 5,280 feet.

NAVSTAR GPS - The Navigation Satellite Timing and Ranging GPS.

P-code - The Precise code. A very long sequence of pseudo random binary biphase modulations on the GPS carrier at a chip rate of 10.23 MHz which repeats about every 267 days. Each one week segment of this code is unique to one GPS satellite and is reset each week.

Precise Positioning Service (PPS) - The most accurate dynamic positioning possible with standard GPS, based on the dual frequency P-code and no SA.

Pseudolite - A ground-based differential GPS receiver which transmits a signal like that of an actual GPS satellite, and can be used for ranging.

RTK - Real Time Kinematic.

Satellite Constellation - The arrangement in space of a set of satellites.

Selective Availability (SA) - A policy adopted by the Department of Defense to introduce some intentional clock noise into the GPS satellite signals thereby degrading their accuracy for civilian users.

Space segment - The part of the whole GPS system that is in space, i.e. the satellites.

Standard Positioning Service (SPS) - The normal civilian positioning accuracy obtained by using the single frequency C/A code.

User segment - The part of the whole GPS system that includes the receivers of GPS signals.

□□□

Colour Plates

Chapter 1: Introduction to Remote Sensing

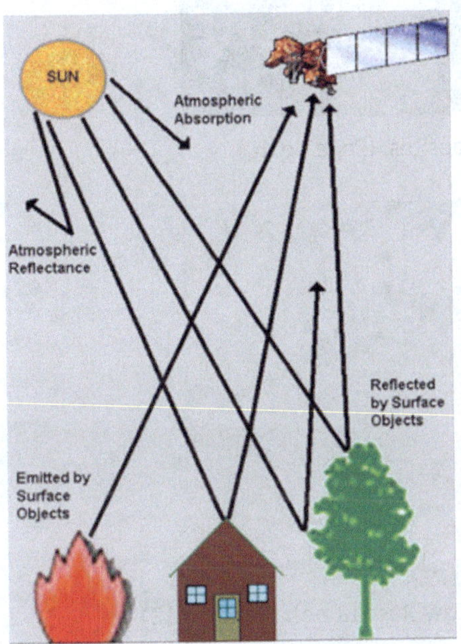

Figure 1.1. Passive System (Page no 6)

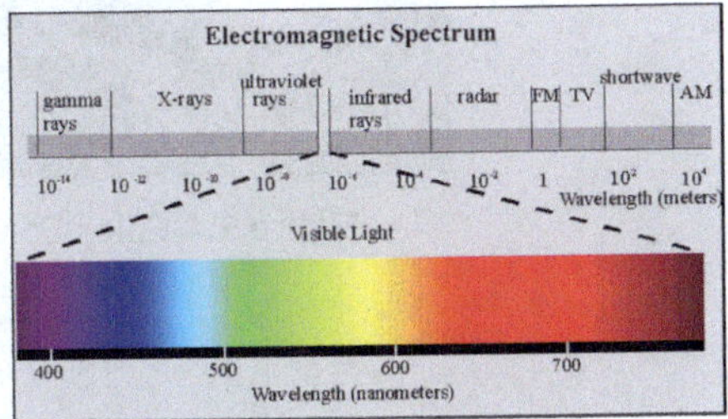

Figure 1.2 Electromagnetic (EM) Spectrum. (Page no 7)

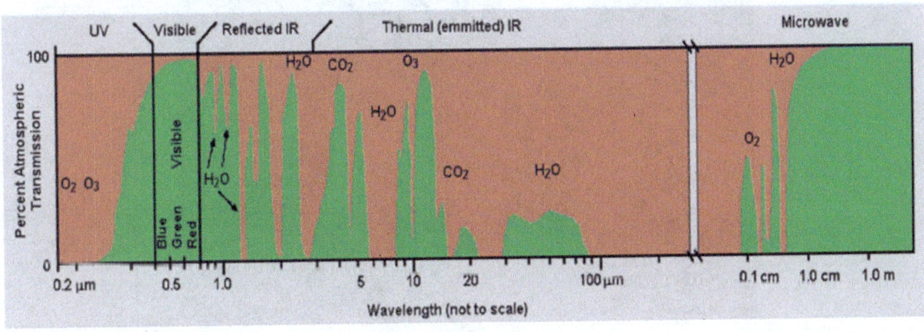

Figure 1.3 Atmospheric Windows. (Page no 8)

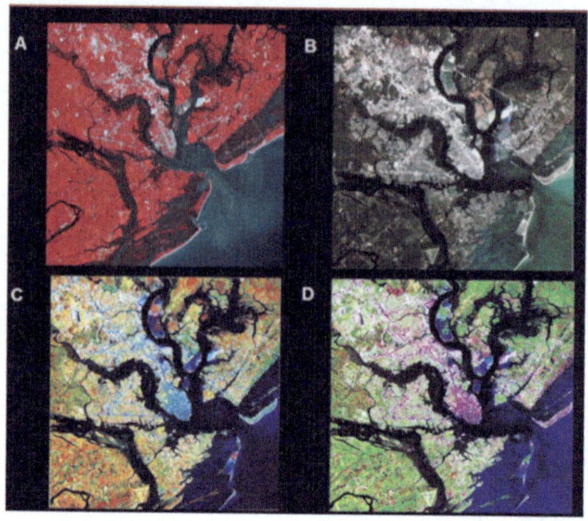

Figure 1.4 Color Composites. (Page no 10)

Figure 1.6 African Dust Storm (Low Resolution). (Page no 12)

Figure 1.7 Madurai Kamaraj University (High Resolution). (Page no 13)

Chapter 3: Satellite Remote Sensing

Figure 3.3 FCC Colour Image (Page no 54)

Colour Composites : Spectral

'False Colour' composite (FCC)

IR band on Red

Red band on Green

Green band on Blue

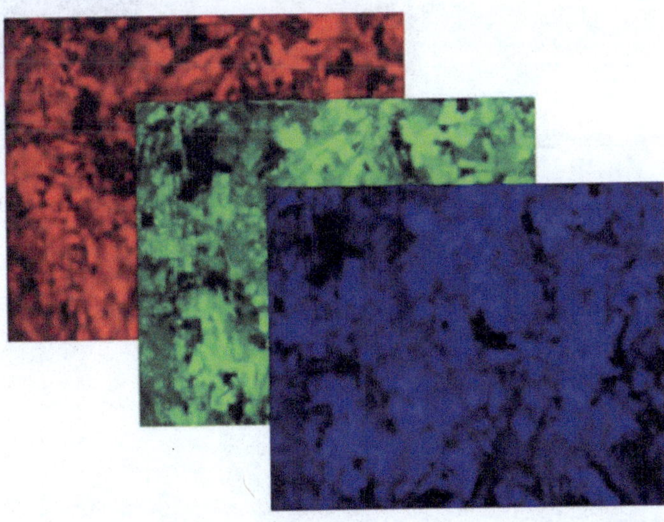

Figure 3.4 Colour composits (Page no 55)

Chapter 4: Global Positioning System

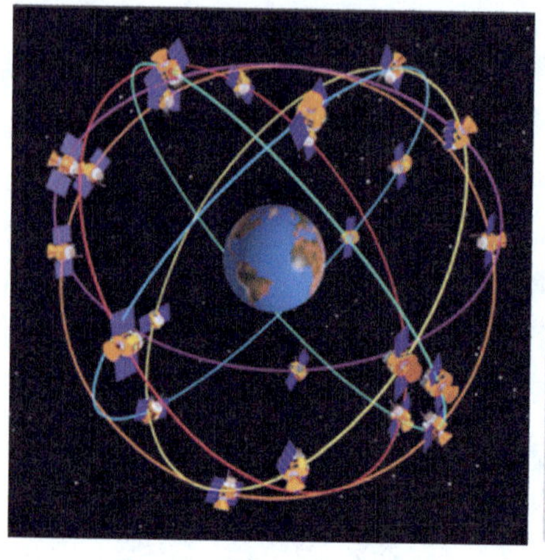

Figure 4.1 Satellite Orbits (Page no 82)

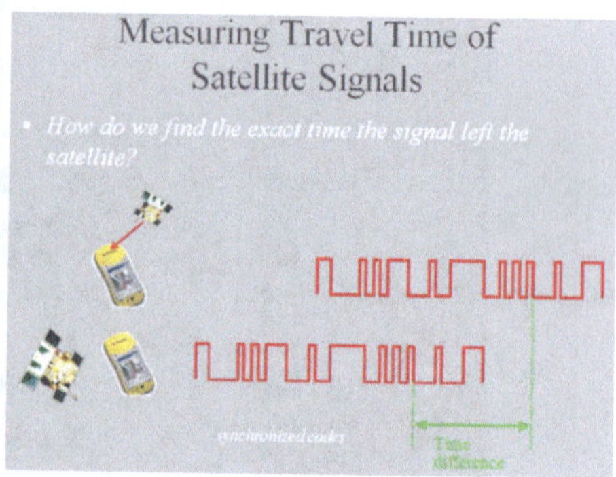

Figure 4.2 Measuring GPS (Page no 87)

Chapter 5: Introduction to Geographical Information System

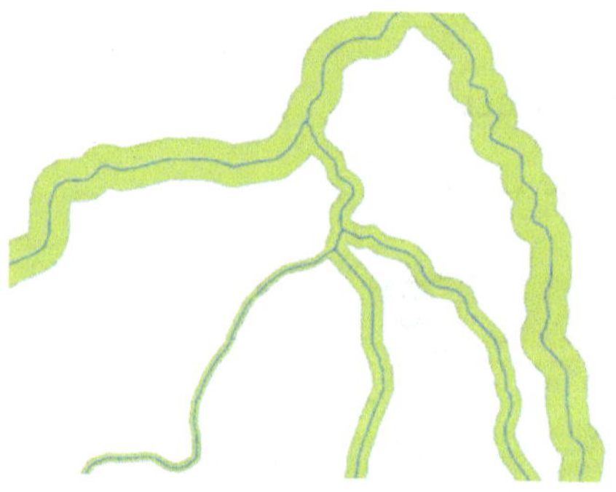

Figure 5.10 Line Buffer (Page no 119)

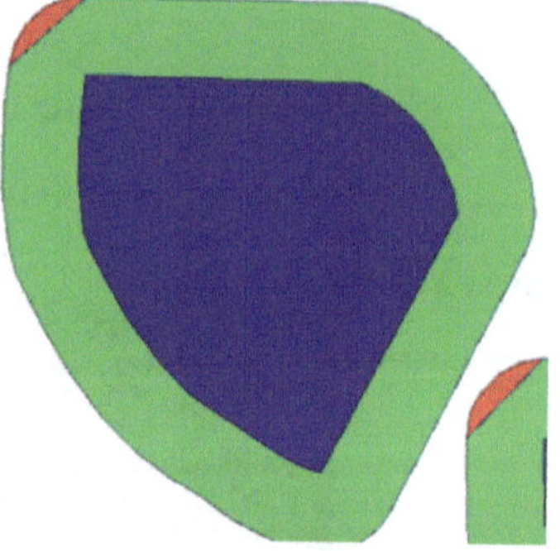

Figure 5.11 Area Buffer (Page no 119)

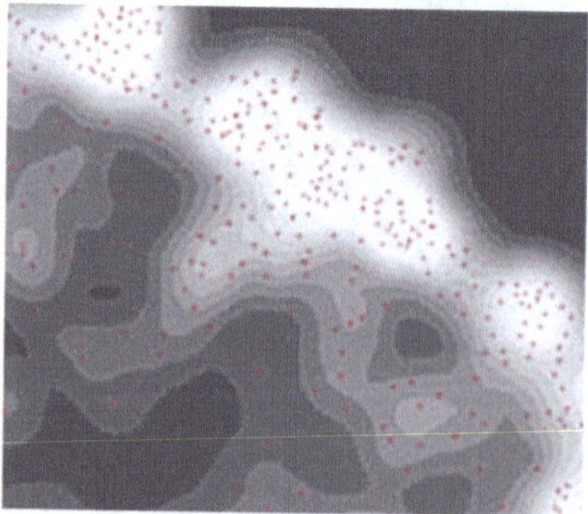

Figure 5.12 Neighbourhood operations (Page no 119)

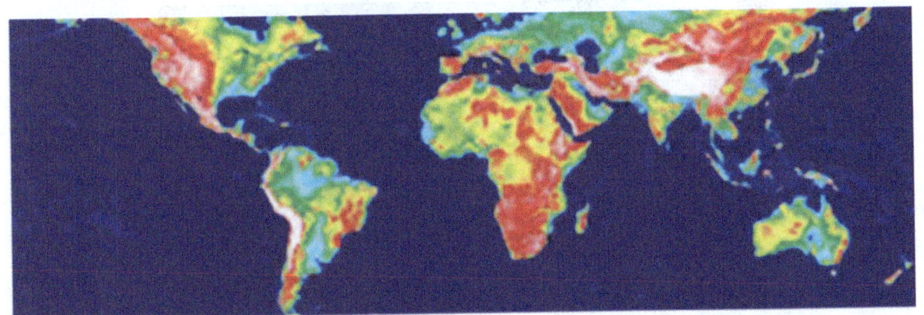

Figure 5.14 SRTM data (Page no 124)

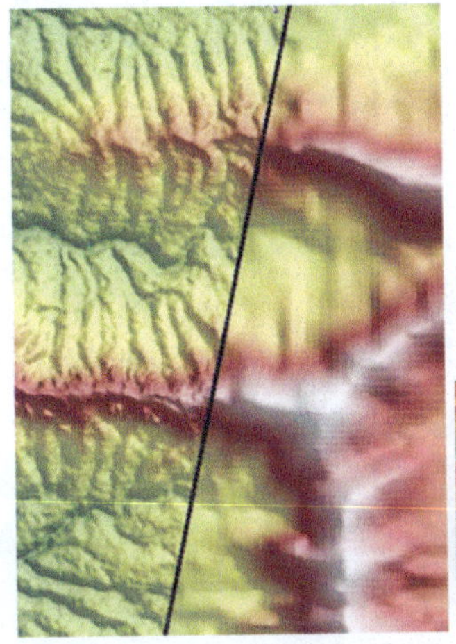

Figure 5.15 ALOS PRISM and SRTM data Comparison (Page no 125)

Figure 5.16 DEM Model (Page no 125)

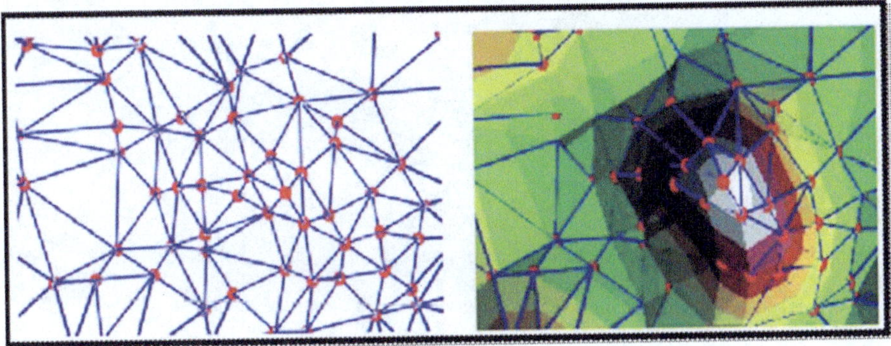

Figure 5.17 TIN Model (Page no 126)

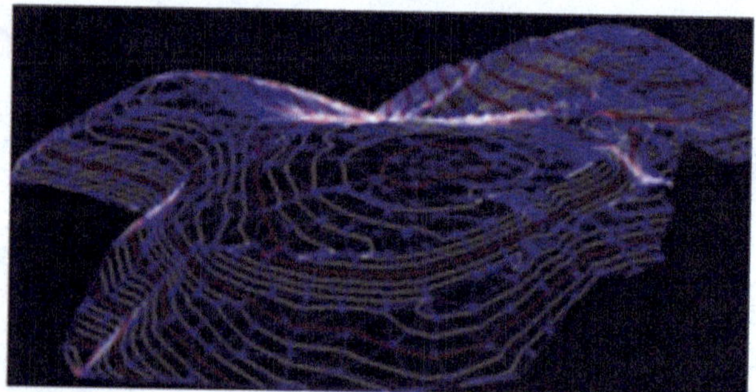

Figure 5.18 TIN overlaid with contour lines (Page no 126)

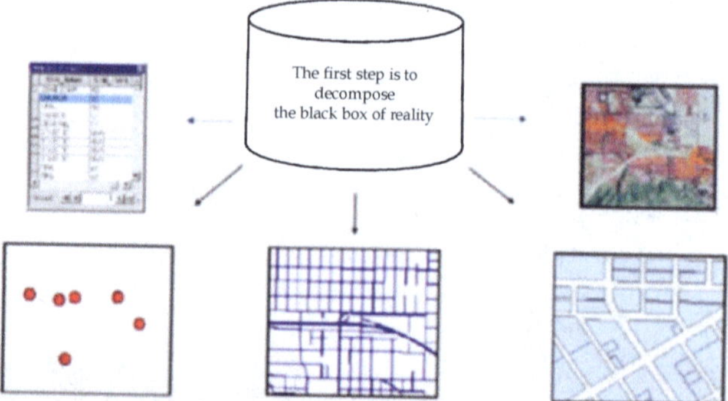

Figure 5.20 store street data (*Source*: Guszlev A. : EDUGI course material) (Page no 129)

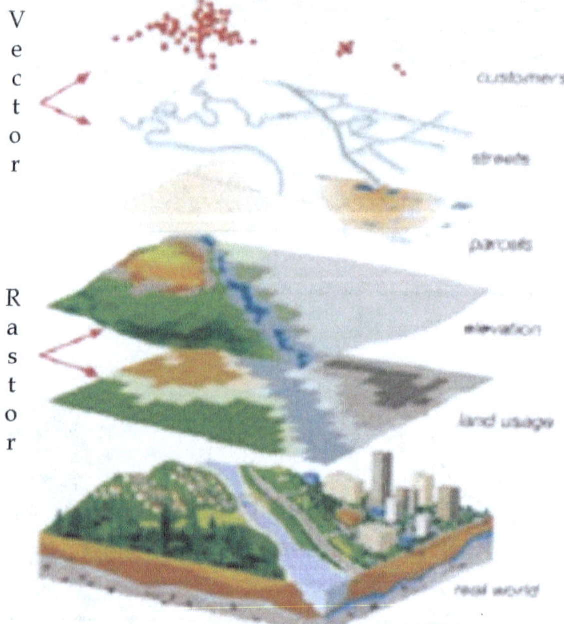

Figure 5.21 GIS Layer (Page no 129)

Chapter 6: Applications of Remote Sensing GIS and GPS

Figure 6.1 Some species (red color) of cultivated plant on satellite image (Page no 133)

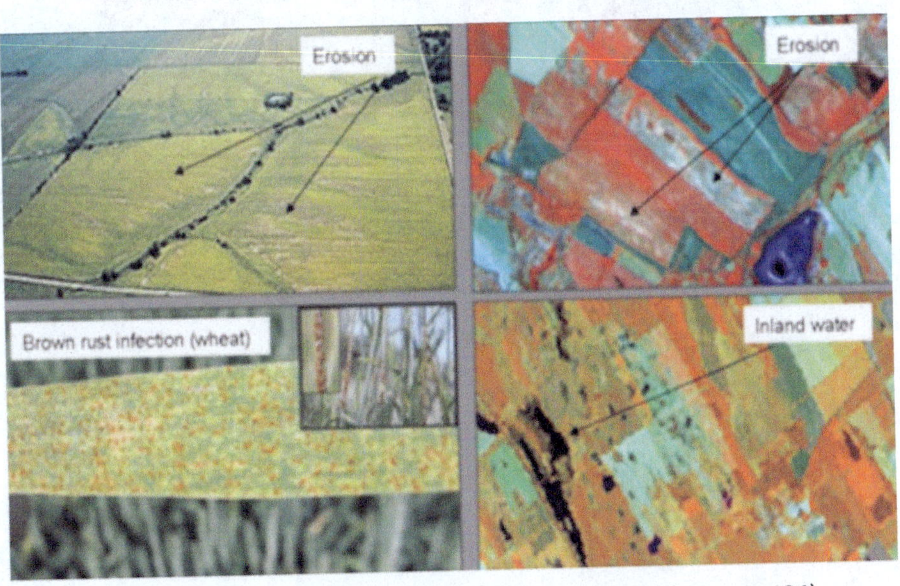

Figure 6.2 The problems inside the agricultural fields (Page no 134)

Figure 6.4 Forest cuts monitoring (LANDSAT TM 1992 and 2000). (Page no 136)
Source: http://www.hso.hu/cgi-bin/page.php?page=84)

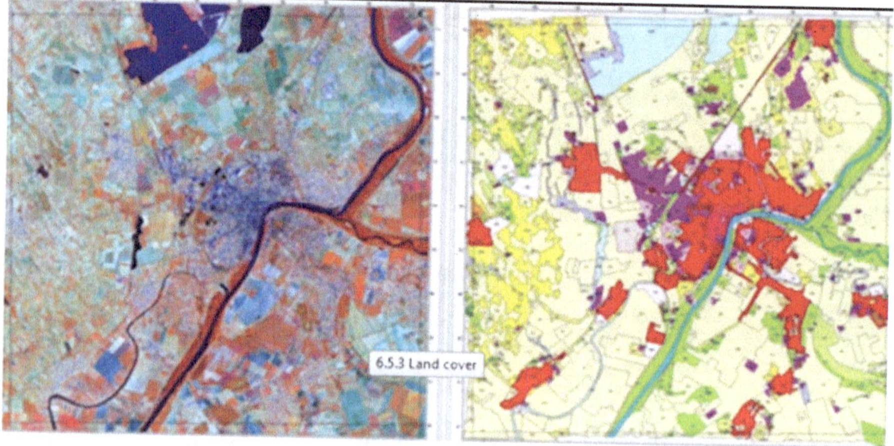

Figure 6.5 Spot satellite images as input data and the result of its interpretation. (Page no 138)
(*Source:* http://www.fomi.hu/corine/clc50.htm)

Bibliography

Jenson, John R. (2007). Remote Sensing of the Environment: An Earth Resource Perspective. Pearson Prentice Hall.

Lillesand, T.M. and Kiefer, R.(1993). Remote Sensing and Image Interpretation. Third Edition, John Wiley, New York.

Lillesand, Thomas M. and Ralph W. Kiefer. (1994). Remote Sensing and Image Interpretation. John Wiley and Sons, Inc. New York.

Sabins Jr., Floyd F, (1978). Remote Sensing: Principles and Interpretation. W. H, Freeman and Company. New York.

Jenson S, Domingue J (1988). Extracting topographic structure from digital elevation data for geographic information system analysis. *Photogrammetric Engineering and Remote Sensing*, 54: 1593-600.

Anji Reddy M (2008). Textbook of Remote Sensing and Geographical Information Systems. Third Edition, BS Publications. New Delhi.

Burrough P A, McDonnell R (1998). Principles of Geographical Information Systems. Oxford University Press, Oxford.

Heywood D I, Oliver J, Tomlinson S (1995). Building an exploratory multi- criteria modelling environment for spatial decision support. *In:* Fisher P (ed.) *Innovations in GIS 2*. Taylor and Francis, London.

Rafael C. Gonzalez and Richard E. Woods (2008). Digital Image Processing, Library of Congress Cataloging-in-Publication Data on File, Pearson Prentice Hall, Pearson Education, Inc. Upper Saddle River, New Jersey.

S. Suganthi and K. Srinivasan (2010). Digital Elevation Model Generation and Its Application In Landslide Studies Using Cartosat1, *International Journal of Geomatics and Geosciences* Volume 1, No 1, 2010.

Molly Reif et.al (2012). Ground Truth Sampling to Support Remote Sensing Research and Development: Submersed Aquatic Vegetation Species Discrimination Using an Airborne Hyperspectral / Lidar System. ERDC TN-DOER-E30, January.

Fundamentals of Remote Sensing, A Canada Centre for Remote Sensing. Remote Sensing Tutorial.

Shefali Aggarwal, Principles of Remote Sensing, Photogrammetry and Remote Sensing Division. Indian Institute of Remote Sensing, Dehra Dun.

Sabins, F.F. (1997). Remote Sensing and Principles and Image Interpretation. WH Freeman, New York.

Colwell, R.N. (Ed.) (1983). Manual of Remote Sensing. Second Edition. Vol I: Theory, Instruments and Techniques. American Society of Photogrammetry and Remote Sensing ASPRS, Falls Church.

Joseph, G. (1996). Imaging Sensors. Remote Sensing Reviews.

Web References

http://www.isro.gov.in/applications

http://earthobservatory.nasa.gov/Features/RemoteSensing/

http://www.seos-project.eu/modules/remotesensing/remotesensing-c00-p01.html

http://www.aboveitallphoto.com/types-of-aerial-photography

http://www.aboveallphoto.com/stl/aerial.html

http://www.4orienteering.com/aerial_photos/30/

http://www.nrcan.gc.ca/earth-sciences/geomatics/satellite-imagery-air-photos/satellite-imagery-products/educational-resources/9291

http://libinfo.uark.edu/gis/tutorial.asp

http://userpages.umbc.edu/~tbenja1/umbc7/santabar/vol1/lec2/2-3.html

http://www.montana.edu/gps/understd.html

https://www.ja-gps.com.au/what-is-gps.aspx

https://en.wikipedia.org/wiki/Global_Positioning_System

www.essex-countynj.org/GIS.pdf

http://www.rst2.edu/ties/gentools/comp_gis.html

http://ces.iisc.ernet.in/energy/monograph1/Gispage3.html

http://desktop.arcgis.com/en/arcmap/10.3/map/working-with-layers/what-is-a-query-layer-.htm

http://www.sli.unimelb.edu.au/gisweb/BuffersModule/BuffSelect.htm

http://en.mimi.hu/gis/buffer.html

http://planet.botany.uwc.ac.za/nisl/GIS/GIS_primer/page_36.htm

http://www.geoimage.com.au/services/dem-generation

https://en.wikipedia.org/wiki/Triangulated_irregular_network

planet.botany.uwc.ac.za

http://www.tankonyvtar.hu/en/tartalom/tamop425/0027_DAI7/ch01s02.html

http://www.tankonyvtar.hu/en/tartalom/tamop425/0027_DAI6/ch01s05.html

http://www.environmentalscience.org/principles-applications-gis

http://www.supergeotek.com/library_GISApplication.aspx

http://grindgis.com/blog/gis-applications-uses

http://www.mio.com/technology-applications-of-gps.htm

http://infohost.nmt.edu/~mreece/gps/conclusion.html

https://en.wikipedia.org/wiki/Global_Positioning_System

http://www.gps.gov/applications/

http://www.unesco.org/csi/pub/source/rs15.htm

http://www.isro.gov.in/irnss-programme

https://en.wikipedia.org/wiki/Indian_Regional_Navigation_Satellite_System

http://www.photogrammetry.com/

http://www.geospatialworld.net/article/achievements-in-remote-sensing-applications-in-india/

https://en.wikipedia.org/wiki/Indian_Remote_Sensing

http://tycho.usno.navy.mil/gpscurr.html

http://www.trimble.com/gps/puttinggps/gpsfram1.htm

http://www.aero.org/publications/GPSPRIMER/GPSElements.html

http://www.colorado.Edu/geography/gcraft/notes/gps/gps_f.html

http://www.rand.org/publications/RB/RB1501/RB1501.html

❑❑❑

S.No.	Title	Author	ISBN	Year
30	Principles and Applications of Agricultural Meteorology	Patra, Alok Kumar	9789385516245	2016
31	Remote Sensing Applications in Dryland Natural Resource Management	Gaur, Mahesh	9789381450321	2013
32	Remote Sensing in Geomorphology	Ramasamy, SM: ed.	9788189422059	2005
33	Town Planning Regeneration of Cities	Ashutosh Joshi	9788189422820	2008